Time for Math!

Creating a Learning Environment That Works for All Students

A TEACHER RESOURCE for K-2 MATHEMATICS

Kathy Richardson

Math Perspectives, Bellingham, WA

Math Perspectives

Editor and Production Coordinator: JoEllen Key

Cover Design: Katrina Lyon

Design and Layout: Katrina Lyon

Illustrations: Linda Starr

This book is published by:
Math Perspectives Teacher Development Center
PO Box 29418
Bellingham, WA 98228-9418
Phone: (360) 715-2782
www.mathperspectives.com

ISBN 978-0-9848381-8-9

Acknowledgements

Writing this book caused me to take a journey back through time as I reflected on all my years in the classroom. I remember so clearly that first year of teaching, when I had 36 children in my first grade room. What a challenging year that was, with nothing quite the way I had imagined it when I was taking college classes. I hope my children learned something that year; I know I did.

It's funny what you remember. My first week of teaching is forever imprinted in my mind. There was Donnie who cried because he didn't have his money for ice cream. I wasn't sure I knew what to do if someone had stolen his nickel, but when I asked what happened, he tearfully confessed he had swallowed it. And I remember the lunch box dripping white foamy liquid and Tammi's surprise and disappointment that the ice cream bar she had saved to eat after school had melted. I also remember the first day I taught Kindergarten. All I said was, "I need some helpers to pass out the milk," and I ended up with 33 children in my lap! It was those little unexpected things that often got in the way of the work I was trying so hard to do.

I recall other kinds of problems as well–the problems that arose because I didn't know how to make mathematics meaningful to my children. I remember Marie, strong and tall and so confident on the playground. But Marie had no confidence in her ability to make sense of the math we were doing. That was very clear to me one day when I asked what I thought was an obvious question: "Is the orange rod longer or shorter than the yellow rod?" The evidence was in front of her, but she didn't know what to say. She was looking for the trick, the right answer, the clue on my face.

And I never will forget Joey. One day, he was the first one done with his worksheet–because he had written "1" as the answer for every problem on the page. When I impatiently asked, "Joey, what are you doing?" he looked up at me and said, "Why you always mad to me, teacher?" I knew I wasn't "mad to" Joey. But I was frustrated with my own teaching and not knowing how to help him. Why was mathematics so hard for my children? They seemed to know how to do something one day and then forget it the next. There was a level of frustration and confusion in my children that was intolerable to me. And even more intolerable was the attitude, "I don't want to understand. Just show me what to do."

Acknowledgements

This wasn't the way it was supposed to be. There were many times when all I could see were the problems. How could I ever meet all the needs that presented themselves to me? How could I create a positive, effective classroom where we were all glad to come each day to spend precious time together? As the years passed, I kept searching for ways to help my children make sense of the work they were asked to do. What most changed my life as a classroom teacher was meeting and working with Mary Baratta-Lorton. She was the first person who not only understood my questions, but helped me figure out some answers as well. Through my work with Mary, I found we could help children make sense of mathematics. We could meet the range of needs in our classroom. We could teach and the children could learn–with joy and enthusiasm.

I have had lots of opportunities to learn during my life as a teacher. The children taught me what they needed to grow and be successful. Yet, I could not have learned as much or grown as a teacher if it were not for the adults in my life who were always willing to listen and try to help me figure it all out. During those early years in the classroom, Janann Roodzant, Agnes Hinojos, Norma Gonné, Concha Cesena and Laura Sanchez were always there to lend a hand and help me keep my perspective. Margie Gonzalez, Deane Reed, Leslie Salkeld, Jody Walmsley, and Mary Baratta-Lorton made me believe my endless questions were interesting and even important. Later, Deborah Kitchens, Ruth Parker, Linda Gregg, Pat Sherbondy, Chris Oster, Jean Church and Cathy Young were there to challenge my thinking, ask me questions, and give me feedback as I worked to communicate what I had learned. My learning continued when Sheryl Russell and I founded Math Perspectives Teacher Development Center. My Math Perspectives colleagues and instructors continue to challenge and support me. I thank you all for your love, support, wisdom, and encouragement.

Much of what I know I learned the hard way, by finding out what doesn't work as well as what does. I hope the information in this book will help some of you avoid the mistakes I made and stay focused on what really matters in the lives of children. We have an awesome responsibility as we work to find ways to create a learning environment that truly supports the children we are there to serve. This book is dedicated to all the children and all the teachers who lived and learned with me, and to all the teachers who continue to find ways to support children as they grow and learn.

Table of Contents

Table of Contents

1 The Learning Environment

What kind of environment are we aiming for?

The Learning Environment

Our goal is to create an environment for learning mathematics in which the tasks engage the thinking of each child; the children are expected to work hard and to make responsible choices; we see a high level of commitment, persistence and independence in each child; and all children are challenged and working at the level appropriate for them.

Mathematics is much more than what many of us experienced as children. Math is all around us and can serve as a tool for exploring and making sense of the world. Mathematics helps us understand and analyze our experiences. We use mathematics to gather and organize data in order to solve problems. It can answer questions like: What do we notice when we weigh our baby hamster as it grows? What happens to the weight of our pumpkin as it decays? What time of day does the most traffic go by our school? Did our plant grow the same amount every day? How many lemons do we need to make lemonade for our class?

But for many children, mathematics is not a subject that opens up the world to them. Rather, it is only something to be memorized and rarely understood. As mathematics becomes an increasingly important part of the elementary curriculum, it can turn into a stumbling block rather than a stepping-stone to future success. But it does not have to be this way. Learning mathematics can be exciting. We can support all children in developing an understanding of math. We present here realistic and practical ways to create an environment where we can meet the needs of children wherever they are in their development of math concepts and help all children be successful.

What Kind of Environment Are We Aiming For?

This is a book that honors and celebrates the way young children learn. It is based on the belief that we can maximize children's learning if we look honestly and respectfully at their way of working and establish an environment that is consistent with how they learn. Children are very good at saying words or following procedures before they understand them. Too often we make assumptions about what they really know and understand, and in the process, we ignore their real needs. We frequently expect either too much or not enough. However, children thrive in an environment that supports them in their search for sense and meaning and provides them with basic concepts and understandings that help build a strong foundation for the learning they will do in the future. The more we know and understand about the real work of the young child, the more we will be able to create a positive learning environment that provides them with appropriate challenges and expectations.

Those of us who work with young children know how important the early years are for building a base for all subsequent mathematical learning. For children to be successful in mathematics beyond the primary grades, they must leave with more than knowledge. They must also develop positive attitudes and habits of mind that cause them to be interested, curious, and eager to learn new mathematical ideas. Children need to see mathematics as something that is useful to them when they solve problems or make decisions. *Time for Math! Creating a Learning Environment That Works for All Students* was written to help teachers of primary children establish a classroom that is supportive of an active, meaning-based approach to the teaching and learning of mathematics that allows all children to learn and all children to be successful.

The learning environment we are aiming for is one that:

- encourages thoughtfulness
- engages children's thinking
- provokes questions
- stimulates a search for meaning
- encourages children to look for connections and relationships
- helps children make sense of and understand the mathematics in which they are engaged

This kind of environment is both realistic and attainable. We can create this type of classroom, but it does not happen instantly. It develops over time and is influenced by all the decisions we make, large and small. In this book, we will take a look at the many decisions teachers must make and the ways these decisions impact the teachers' ability to establish the desired environment in the classroom.

When we consider what it takes to create a positive learning environment, we need to think not only about the physical layout of the room, but we must also reflect on the settings in which children work, the kinds of tasks in which they are engaged, and the way we interact with our children. All these factors must be taken into account as we work to create a supportive learning environment that meets the needs of all our students.

To create this environment, we may have to give up some of the ways we have managed our classrooms in the past and judge our success in ways that might be new to some of us. We will find that the quiet classroom is not always the most productive classroom, but neither is a classroom that has activity for the sake of activity. Some of you will find it challenging to allow noise and movement as children work. Others will feel anxious when asked to have children experience the same set of tasks over time when doing so will result in using fewer of all those "good ideas" we are so eager to share with our children.

If the approach to teaching and learning presented here is new to you, you are bound to encounter some problems and frustrations. We know that things do not always go smoothly when working with children. It is not necessary that everything go exactly as anticipated. What is necessary is a willingness to try and a willingness to be thoughtful and reflective about those things that go well and those that don't. I am confident that if you put these ideas into practice, you will find math time one of the most rewarding times of the day.

2 Mathematics for All Children

What mathematics should children be learning?

How do children develop an understanding of number concepts?

What kinds of tasks support the development of concepts?

How can we use tasks to meet the range of needs in our classroom?

Mathematics for All Children

Our work is to build a strong foundation in primary grades, enabling children to make sense of their work with mathematics.

What Mathematics Should Children Be Learning?

Mathematics is more than the study of numbers and how they work. Children also need experiences with geometry, measurement, collecting and organizing data. Even so, the development of number concepts is at the heart of the mathematics program for young children. We should not underestimate the importance of building a strong foundation in number. What children know and understand about number impacts their work with all other mathematical topics. Children need ongoing and multiple opportunities to develop number sense: to count and compare quantities, to add and subtract, to work with place value in ways that ask them to think and reason, and to see relationships and make connections.

No matter what textbooks or math programs are adopted, there will always be certain foundational number concepts that should and will be addressed. These core concepts are:

- Counting
- Comparing Numbers
- Addition and Subtraction
- Place Value
- Multi-Digit Addition and Subtraction
- Multiplication and Division

When planning for a year of mathematics, I consider number concepts to be the core of the program and so I establish ongoing work with number as the basis of the children's work. Other mathematical ideas lend themselves to a natural integration with this work with number. For example, measuring is a great context for developing number concepts and relationships. Data collection and graphing, which are ongoing activities woven into the math program all year, also support the development of number relationships. When children use dice in playing games, they have the opportunity to begin working with ideas of probability as they notice whether some numbers come up more often than others. Other major areas of emphasis, such as geometry, pattern, and sorting can be interspersed for a week or two at a time throughout the year.

Standards and goals are important in establishing what needs to be taught. However, it is important to acknowledge that children will reach these standards at different times. The achievement of any particular standard comes at the end of a long process and cannot be taught without consideration of the underlying ideas necessary for understanding. For example, a goal for a group of children may be to develop proficiency adding and subtracting multi-digit numbers, but individual children within the group can be working towards this goal at the level that is appropriate for each of them. One child might be learning to make one ten and some left over ones and to find the total without counting, while another child is adding groups of tens and ones. Even though their progress may be at different places, both are working towards the achievement of the same goal.

When we move all children forward at the same time, we run the risk of skipping the foundational experiences that some of the children still need. This results in ongoing confusion for these children and requires a continuous review of what should have been learned in previous years. A case in point: Teachers of children in 5th grade and beyond commonly report that many of their students struggle with place value concepts. This lack of understanding makes it very difficult for students to work successfully with more challenging computation problems. Concepts like decimals that should build on these understandings are mysterious and frustrating. The students' incomprehension isn't because they haven't been "taught" place value in the past. Rather, it occurs when they have been asked to work with place value concepts in ways that did not help them to develop understanding, or were asked to work with concepts they were not yet ready to understand. We cannot assume that children are learning place value concepts simply because they can use the language of tens and ones or follow procedures correctly. We need to know how they are thinking about the relationships inherent in the place value system. This simply will not happen at the same time for all students.

How Do Children Develop an Understanding of Number Concepts?

No matter what standards are put in place or what expectations for grade levels are set, children develop an understanding of number concepts in predictable ways. There are stages of thinking that children go through as they develop understanding and competence with mathematical ideas. I have identified these stages of learning as Critical Learning Phases. The Critical Learning Phases are those levels of thinking about mathematical ideas that are milestones or hurdles in children's growth of understanding. The Critical Learning Phase that a child has reached determines what kinds of instruction will be meaningful and effective. Instruction which is given before a child understands the foundational ideas necessary to make sense of the mathematics is not only ineffective but often leads to misconceptions. Teachers must take the development of children's thinking into consideration when deciding what kinds of experiences individual children need.

Children generally move from thinking of numbers as a collection of single units (that is, one and another one and another one) to seeing that one number is contained within another number, then to learning the parts of numbers to 6 and then to 10. The next big milestone is recognizing that numbers are composed of tens and ones and eventually understanding numbers as hundreds, tens and ones. At every stage of development, the size of the numbers and the size of the differences between numbers influences what the child is able to understand. That is, a child may be able to see that 5 is contained in 7 but not yet know how 15 is related to 17. Or a child may know the parts of numbers to 4 but not yet be able to figure out the parts of 6.

Each stage of learning is much more complex than is generally recognized. If we look at counting for example, we will see that children who are competent counters have integrated several major ideas about counting. These include one-to-one correspondence, inclusion, keeping track, attending to how many they have counted as they go, and knowing one more and one less. A child who understands counting will be able to keep track of an unorganized group of objects and will also be able to count out a particular quantity from a larger group of objects.

The stages that children move through on their way to understanding numbers are described more fully in *How Children Learn Number Concepts: A Guide to the Critical Learning Phases* listed in the References. Here are two examples of stages children move through on their way to understanding particular Critical Learning Phases.

The following example shows the stages related to just one aspect of counting: the ability to count out a specific number. These stages are natural levels of thinking that children go through while they are learning a concept.

Picks up the one "named," the number asked for	Counts past the number asked for, doesn't notice	Counts past the number asked for but self corrects	Gets a certain quantity when asked
Sally's mother asks her to get three spoons from the table. Sally carefully counts the spoons"1, 2, 3" and then picks up the spoon she touched when she said "three." Here is three," she says as she hands the spoon to Mom. ▪ Sally understands counting as labeling each spoon.	Donny's teacher asks him to put 4 scissors in the scissor can. Donny begins to count the scissors as he fills the can. "1, 2, 3, 4, 5, 6, 7 . . ." He continues to count and place scissors in the can, not noticing he has passed the required number. ▪ Donny is not yet able to hold a number in mind as he counts. He is thinking of one and another one and another one rather than thinking of the quantity as he goes.	"I am going to get six napkins for my table," Jenny tells her friend. She stacks the napkins carefully as she counts, "1, 2, 3, 4, 5, 6, 7, 8. Oops – I got too many. I have to start over." ▪ Jenny knows she wants to count out a specific number of napkins, but is not aware of how many she has as she goes. When she hears herself say the number, then she realizes she has gone past.	Jill has been asked to get 12 pieces of paper for the children at her table. She confidently counts out the 12 pieces of paper and delivers them to the children. ▪ Jill can keep the number in mind as she counts and is aware of the numbers she has counted along the way, so is able to stop when she reaches the desired number.

Mathematics for All Children

Another Critical Learning Phase relates to children's understanding of numbers as tens and ones. While children are often able to label the digits in a number as tens or ones, that does not insure that they understand the underlying structure of the numbers or that they know the total number in a group of objects once they know the number of tens and ones. The following stages show what children do to determine totals on their way to understanding this idea.

Counts objects without organizing into tens	Organizes into tens and ones but counts all	Organizes into tens and ones, counts by tens and then counts the ones	Knows total instantly when number of tens and ones is known
Jo has been assigned the task of determining how many tiles will fit on a Paper Shape. She has been asked to find out how many tens and ones it will take to cover the shape. She uses two colors as suggested but does not organize the tiles in any way. When asked to tell how many, she counts all the tiles by ones. ▪ Jo is thinking of numbers as one and one and one and is not yet able to organize into groups.	Jue May is following directions when she covers a Paper Shape with tiles. She counts the tiles as she goes, and each time she gets to ten she chooses a different color. When she is finished, she reports that she made 3 tens and 7 leftovers. When asked to tell how many altogether, she ignores the groups of ten and counts each tile. ▪ Jue May understands the task but does not see that organizing into tens is helpful at all in determining the quantity.	Javier covers a Paper Shape with tiles. Each time he has ten, he changes to another color. He reports that he could fill the shape with 4 tens and 6 ones. When asked to tell how many he counts "10, 20, 30, 40, 41, 42, 43, 44, 45, 46." "There's 46," he says. ▪ Javier can work with groups of objects and counts by those groups to determine the total quantity. However he is not yet able to count the groups and use that to help him know the total.	Barbara covers a Paper Shape with blue and yellow tiles, organizing them into groups of ten as she goes. When she finishes, she says, "There are 4 groups of ten and 3 leftovers. That's 43." ▪ Barbara is able to count the groups of ten and recognizes that the number of tens and the number of ones makes the total.

When children understand that numbers are composed of tens and ones, they can learn to combine the tens and ones with relative ease, forming additional tens as needed. However, it is not effective to teach a child to "add the tens" if to them the "tens" are simply digits in a number. They may learn to get answers but without understanding. When children receive instruction before they have the foundational ideas necessary to understand the mathematics present in the problems they are asked to solve, the best they can do is memorize the steps for getting right answers to these problems.

In addition to the size of the numbers, the level of abstraction also effects what children are able to think about. As children develop their understanding of number, they begin with models, actually moving them to aid their thinking. They generally advance from being able to think about numbers when they can move objects–to thinking about relationships when the model is present but not touched–to thinking about relationships without a physical model. Only after children can work with an idea at all these levels will their experiences with the symbolic representations of this idea be easy for them for them to understand. Even then, young children rarely reach the level of thinking that is totally abstract. When working with symbols, they usually need to be thinking of some concrete referent. That is, they need to think of 3 tens as 3 trains of Unifix cubes or see 3 + 4 as 3 cookies and 4 cookies.

The most efficient way for a teacher to support a child's efforts to meet a standard they have not yet achieved is to begin wherever the child is and have them work towards whatever the standard requires. Doing anything other than this simply wastes the child's time and prevents them from developing the essential foundational skills for future success.

What Kinds of Tasks Support the Development of Concepts?

The most important thing about a mathematical task is that it requires the child to think. Tasks do not have to be complicated or elaborate to engage children in thinking. When we understand and respect the complexity of the seemingly simple work young children do, we will see that what may seem simple to us is not really easy at all. In fact, it is often the tasks that appear to be simplest that lead to the most focused attention on the mathematics involved.

Mathematics for All Children

To be most effective, the tasks must make sense to the children and be simple enough so that learning the directions doesn't get in the way. We want the children to focus on understanding the concepts and not get lost in the instructions for doing the task. The most powerful tasks have value in being repeated over and over again. Each time children do a task they will be working at a different level as they develop understanding and increase competence.

Consider the following activity, "Creating Shapes on the Geoboard," from *Understanding Geometry* (listed in the references) where the question is posed: "How many different seven-sided figures can you make on the geoboard?"

The directions for doing this task are easily understood by children, so in that sense it is a simple task. However, actually doing the task can be quite challenging and complex. If we watch children at work, we will see their surprise when they think they have added a side only to find out they have ended up with the same number they started with, or that they added two sides instead of one. When children first work with this task, they count and recount to check the number of sides. Then, with experience, they begin to be more deliberate and eventually, develop strategies for adding or taking away sides. This increasing competence can develop only if they have numerous opportunities to work with this task.

Measuring yarn using paper clips and Unifix cubes as the units of measure is another simple task that requires complex thinking on the part of the child. While measuring the yarn, the children can:

- Estimate. *"I think it will take 23 paper clips."*
- Check and revise their estimates. *"I've already used 15 paper clips and there's lots of yarn left to measure. I'm changing my mind. Now I think it will take 35 paper clips."*
- Organize into tens and ones. *"Every time I get to 10 paper clips, I mark it by putting another paper clip next to it. Then I can tell how many I have without counting each one."*
- Record numerals. *"I counted 32 paper clips. I can write that number down."*
- See relationships. *"I used 32 paper clips. The Unifix cubes are smaller. I think it will take about 40 Unifix cubes."*
- Compare. *"It took 32 paper clips and 59 Unifix cubes. It took 27 more Unifix cubes than paper clips."*

There are many advantages to using simple tasks with children. We are not spending a lot of time preparing or organizing material or teaching children complicated procedures. We will then be able to focus on the interactions that are such an important part of the teaching and learning process.

Mathematics for All Children

How Can We Use Tasks to Meet the Range of Needs in Our Classrooms?

A classroom is a place where a group of unique individuals come together to learn. We know that these individuals do not all learn the same thing at the same time even when they have the same experiences. The development of mathematical understanding and competence does not happen on the same time line for all children in a particular grade level or classroom. What children learn depends in large part on the knowledge they bring to the experience and their cognitive maturity. Let's consider a common goal for kindergarten: Counts and keeps track of up to 20 objects. In any kindergarten class there will be some children who will already know how to count and keep track of 20 or more objects even before receiving any instruction from their teacher. But there will also be children who have not yet learned the rote counting sequence to 10 and will need to learn that sequence before working to learn the sequence to 20. There will be other children who know how to count by rote to 20 but are not yet able to keep track of more than 6 objects. Others may be able to count 20 objects accurately but only with great effort and concentration—who are so focused on the act of counting that they do not attend to the number they arrive at in the end. Still other children may count with ease so are able to think about number relationships and react to what they are counting. They might think something like, "I thought there were 7 but I already counted to 11 and I'm not done yet. It's going to be a really big number."

Expecting all these children to work with the same tasks using the same numbers will be a waste of time for many of the children. Our goal should be for all of our children to work to what I call the "edge of their understanding." That means they all need to be working with tasks that are appropriately challenging for them, no matter what their abilities or experiences. Children should not be doing activities that are out of reach and have no meaning for them – but there also should be no upper limit to what the children can learn, notice, discover, question, or wonder about the tasks they are doing.

Expandable Tasks

Meeting all the individual needs in the classroom is a big challenge. However, if we are careful about the tasks we choose, we can reach this goal. Rather than presenting different tasks to different children, a more manageable approach is to provide tasks that are "expandable" and thus fit the needs of various children depending on how they approach the task. *The Developing Number Concepts Series* is one place to find expandable tasks such as those described here.

Sometimes expandable tasks allow the children to work with different quantities. One child might be building toothpick designs and labeling the parts using 6 toothpicks while another child may be using 8 toothpicks. One child may be counting out 5 apples on each tree while another child is placing 10 oranges on each tree.

Sometimes the task is the same but children are focused on different things. The following is an example of how one task might meet a range of needs depending on how each child thinks about it.

Instead of measuring the length of a piece of yarn, a group of second-graders made various shapes using pieces of yarn. Their job was to figure out how many tiles fit in the shapes. The children were focused on different skills as they worked with the task.

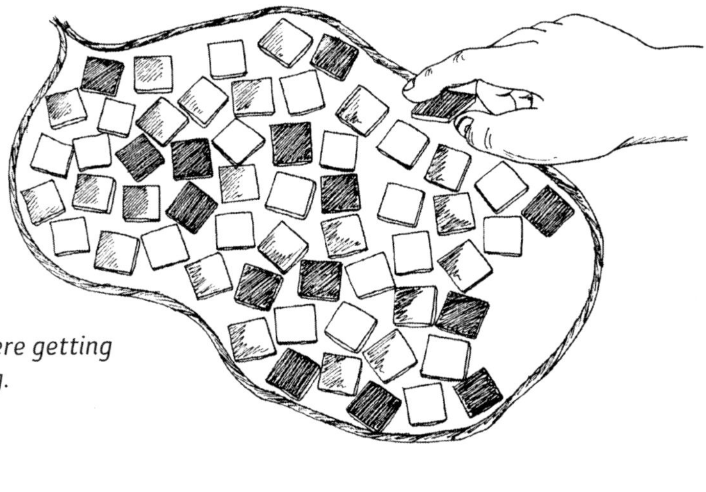

Some children were getting practice counting.

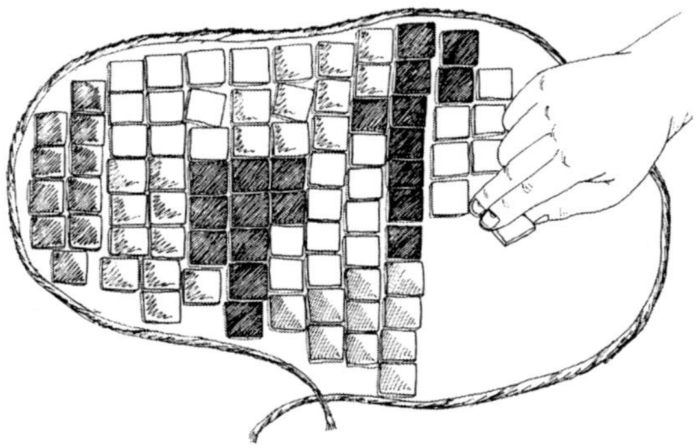

Other children were beginning to develop strategies for organizing to make counting easier, and using color to form groups of ten.

One child was finding a shortcut and filling one-half of the shape and doubling the number instead of counting each tile.

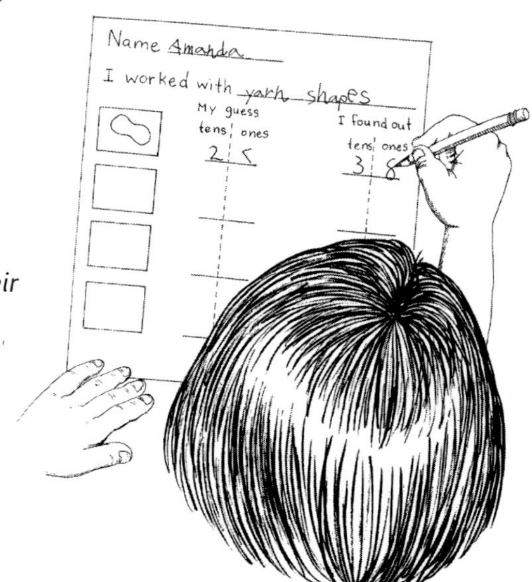

Other children were comparing their estimates to the actual count and figuring out the difference.

While the children were generally focused on number concepts when they worked with the yarn shapes, the task also provided them an opportunity to look informally at the relationships between area and perimeter.

Jolene figured out that she could make a shape that uses a very small number of tiles if she made it long and skinny.

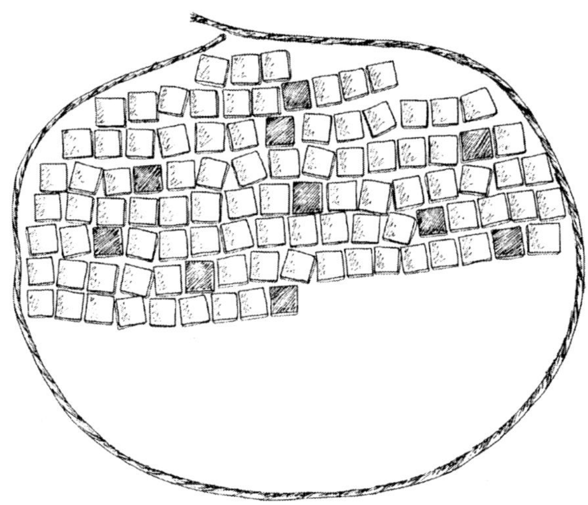

Mikkal wanted to make the biggest shape he could and formed a circle.

Each child is getting something different from the task, but every child is learning and every child is challenged.

We can most effectively meet the range of needs in our class if we allow our students the opportunity to work on problems using their own methods and approaches. To a large extent, when tasks are expandable, children will naturally work at an appropriate level. However, we must also be alert to opportunities to help children gain the most benefit from the tasks. We need to watch and listen as they work, interacting with them and offering support and challenges when needed.

Math in the Real World

Sometimes, appropriate mathematical experiences can be a natural part of the study of many topics and explored through theme-based instruction or via children's literature to provide context for the students. But we need to be careful when selecting mathematical tasks and experiences to be sure that the mathematics is authentic and central to the experience. If the theme is intended to be a study of bears and a context for learning mathematics, then our work with bears should enhance the children's understanding of both bears and mathematics. But an understanding of bears is not enhanced by estimating the number of gummy bears in a jar, or by measuring a teddy bear. What often happens is that the children do not learn either topic well. It's not necessary to have theme-related materials in order to discover relationships and develop a sense of quantity. For example, children can be productively engaged when estimating the number of marbles it takes to fill a jar and then comparing that to the number of pecans it takes to fill it. Using gummy bears is irrelevant and only creates the illusion that there is a meaningful context. If there is no natural real-world connection to the topic, it is better not to try to force a connection.

Another problem with giving children mathematical experiences that are integrated into the study of other topics is that they usually do any particular task only once. One measuring experience will not help children develop an understanding of measurement. Children need to be immersed in the concept of measuring if they are to deepen their understanding. If we choose math tasks because they go with a theme rather than because we want to support the development of a particular mathematical idea, we will not be providing the in-depth work our children need.

Mathematics for All Children

We want children to understand and make sense of the mathematics they are learning. This means the situations in which children work must make sense to them, but it does not mean that these situations must be "real life" applications. Sorting a box of buttons and looking for patterns on the 1-100 chart can be important mathematical experiences for children and do not need a real world context to be meaningful. There is richness, power, and beauty in the study of mathematics itself.

3 Teaching for Understanding

What does it mean to teach for understanding?

How can we teach in ways that uncover what children understand?

How do we deal with children's errors and misconceptions?

Teaching For Understanding

What is obvious to us is not always obvious to children. It is important to know what children don't understand.

What Does It Mean to Teach for Understanding?

How children experience mathematics has a profound effect on **what** they are able to learn. If we want our students to be successful in mathematics, it is vitally important that we create a learning environment that supports the development of understanding. If children learn mathematics in a way that allows them to understand what they are learning, they discover that mathematics makes sense and begin to trust their own capacity to make sense of it. The ability to do mathematics well develops in children as they learn to see relationships and to make connections.

On the other hand, when our focus is on how children perform rather than on their understanding, we tend to teach children simply what to do or what to say. We create illusions of learning that we often view as success but that in fact mask children's lack of understanding. Children who learn merely to follow procedures to get correct answers will not trust their own abilities to make sense of mathematics. They will not look for relationships and will not be able to apply what they know in new situations. If we are to provide the most effective environment for our students, we must consider what it means to teach for understanding and to look closely at our interactions with our children and at our expectations for their learning.

Consider the common practice of asking young children to learn to count by tens. Many of us may have been impressed by the fact that young children are able to learn this. But a closer look reveals that saying the counting sequence is not the same as understanding what it means. There are many situations where children's lack of understanding shows up. For example, when asked to count a group of objects by tens, some children will move the objects one at a time while saying, "10, 20, 30" and so on. Some even go on to proudly announce they can also count them by fives or twos – still moving one at a time for each word of the sequence as they demonstrate their accomplishment.

When children show their misunderstanding of what it means to count by tens, we might assume they simply need to be shown how to do it "right." However, even showing children does not ensure understanding. Over the years, kindergarten teachers have tried to give children meaningful experiences with groups of tens. Many have their students mark each day of school with a straw or a Popsicle stick and make bundles of ten as the straws or sticks are accumulated. The children learn to count along with the teacher and the rest of the class and name the right number of tens and ones when the teacher asks them to. However, if you ask these same children at the end of the year how many they think are in a bundle of straws, you will see that the children have many different ideas of what it might be. They call the bundle "a ten" but think there could be 7 or 9 or 12 straws. One teacher shared her surprise at the children's responses when the bundles of ten had been spilled on the ground. "We had just figured out that we had 8 bundles of ten and 4 extras in our can. The children could easily tell me that that meant we had 84. But that day, the can got knocked over, many of the brittle rubber bands holding the groups of ten broke and Popsicle sticks were all over the floor. I told the children we wanted to be sure we had them all picked up. I asked them how many we should have when we were finished. To my total surprise, the children made all kinds of guesses about how many there could be." Contrary to the teacher's surprise, the children's responses were actually typical for children of kindergarten age. Most do not yet understand what it means to think about numbers in terms of groups.

All of us remember times when we also didn't "get it," even if we were able to perform processes or memorize procedures. We learned to say the right words, but we didn't quite know what they meant. There was a level of insight, an "aha," that was lacking. On the other hand, when we (and children) truly understand a concept, we are not just imitating a process or following a procedure. Understanding means we are making connections and seeing relationships. We "get it."

Teaching for Understanding

If we watch children and ask questions of them to find out what they understand, we become aware that what is obvious to us is not always obvious to children. As we work with children, we find that we cannot force understanding. As is true for all learners, children must construct understanding for themselves. They need to go through their own process of "getting it."

As we consider what it means to teach for understanding, it can be helpful to reflect on two states of mind we have all experienced as learners. Piaget described and labeled these states of mind as: *equilibrium* and *disequilibrium.* When learners are in a state of equilibrium, they believe they understand something and are quite comfortable with what they know: they are not actively thinking or trying to figure anything out. However, when something happens that causes them to be surprised or unsure, they either withdraw from the situation or their brain begins an active process of trying to figure it out. Their mind becomes actively engaged in a search for understanding. When what we are experiencing contradicts our previous understandings, when we feel unsure, puzzled, surprised–even confused, we are in that state of disequilibrium. This is not undesirable; rather, it is the most opportune time for learning. Our brain wants to return to a state of comfort and a sense of knowing and understanding, and so is motivated to make sense of what we are experiencing. We want to figure it out.

The way we interact with children during this process is key. Helping children make sense of concepts requires very different kinds of experiences and interactions than does the kind of teaching that asks children to memorize answers and procedures that have no meaning for them. While we can't force children to understand, we can get them engaged in trying to figure things out by setting up situations and asking questions. The key is to get children thinking.

I am reminded of an experience a teacher once shared with me. She wanted her children to compare jars and determine which jars held more and which held less. She brought in several mustard jars that were the same shape but different sizes. Her students (six- and seven-year-olds) focused on the shape of the jars rather than the size and predicted they would all hold the same amount of rice.

The teacher and the children then proceeded to check out their predictions. They first poured one jar into another to see what would happen. Then they counted the number of scoops it took to fill each of the jars. After that, they poured each of the mustard jars into a calibrated container to measure how much each one held. When all the investigations were finished one of the young students, David, said, "Well, I see it. But I don't believe it!"

This story reveals two important principles about teaching for understanding. First, it points out that even events we experience won't always convince us or make us understand. We see that while we may give children opportunities to develop understanding, we can't assume that they have learned from those experiences. It is another reminder that what is obvious to us is not always obvious to children. It's not what they see, but what they **understand** about what they see that makes the difference. Second, we can see by David's comment that he is trying to make sense of what he is seeing and not just trying to figure out what the right answer is supposed to be. His spontaneous remark reveals that his teacher has encouraged sharing and is interested in what the children are really thinking and understanding. Her students are comfortable enough to be honest about what they do or do not understand and don't feel a need to pretend to understand if they don't.

Teaching for Understanding

Confusion is a natural part of developing understanding. If we look back at our own learning, most of us will remember periods of confusion, of not quite getting it as we worked to figure something out. These periods of confusion or incomplete understanding are not only natural, but sometimes may even be necessary. The fact that children become confused or don't fully understand is not a problem or cause for concern. What can be a problem is **our own lack of awareness** of their misunderstanding. If we assume children comprehend when they don't, for example, if we never ask a child to show us what they know about counting by tens, we will not know to provide the appropriate experiences. On the other hand, when we are aware of children's misconceptions and incomplete understandings, we will not make assumptions about their ability to understand concepts they may not have the maturity or experience to understand. We will then be better able to provide what they need.

What does this tell us about our teaching if even experiences don't get the point across? Does this mean we just let children leave an experience thinking whatever they want to think? Not at all. Rather, it means we have to remember that understanding cannot be a one-lesson objective. Understanding develops and deepens over time. When a child does not understand, it means they need more experiences working with the idea. David was not convinced yet, but this experience was still very important to his learning. He is in a state of disequilibrium and will remain engaged with this problem for as long as it takes to make sense of it. His interest in the problem has been stimulated. He will begin to look at jars more closely and ponder what is going on. He will continue to think. The teacher's responsibility is to make sure he has these opportunities and will continue to confront what he does not yet understand.

Children are much more capable, powerful, and confident when they are allowed to make sense of things instead of trying to follow someone else's way of thinking. However, it is important to recognize that allowing them to make sense of ideas in their own way does not mean we stand by waiting for children to learn. Rather, we must be clear about what we want the children to know and understand, and we must provide every opportunity for them to come to understanding. Because understanding is achieved through the direct personal experiences of the learner, our job is to give students appropriate experiences in which they encounter, interact with, and think about important mathematical ideas. We stimulate the children's thinking about their experiences and focus their attention on the mathematics through posing problems and asking questions.

Kindergarten children can engage in tasks that model what counting by tens actually means. For example, adding a dot to a ten frame each day to keep track of the days of school allows children to engage with the experience at different levels. Children can always count each dot to see how many there are. Other children will learn to count the ten frames without yet realizing they are counting "groups." Others may begin to see the relationship between the number of ten frames and the number of tens. The opportunity is there for all children to make sense of the experience in their own way and to grow in understanding as they begin to "see" more.

How Can We Teach in Ways that Uncover What Children Understand?

Teaching for understanding is a more complex process than teaching procedures; it poses different problems for teachers. When the goal is learning procedures, we can easily tell whether the children have learned to do it correctly or not. Children learn to do a particular task in a particular setting. We take for granted they will be able to apply what they know in other situations. It is easy to assume they know more than they do because they can learn to use words quite effectively without understanding what those words actually mean.

However, when the goal is understanding, we often don't know whether the experiences we have provided for our students have been effective or whether they have not. If we truly want to teach for understanding, we need to know what children really know and understand, not just whether they have learned to follow directions. We need to know what they really think, not just what they believe they are supposed to say. How children think is not always visible to adults, so we need to find ways to uncover the thinking behind the answers.

Teaching for Understanding

Much of the most useful information we acquire about children's thinking is necessarily inferred from their actions and comments made while they are engaged in mathematical tasks. If children are encouraged to make sense of their work for themselves, they will share their own thinking instead of trying to read the teacher's face for a clue to the right answer. We must strive to ask questions that give us insight into children's thinking as well as more carefully observe the strategies that children use to get answers. Our goal is to elicit authentic responses that give us authentic information.

Tasks that can be approached in a variety of ways will reveal more about children's thinking than tasks that must be done in a particular way. Consider the following examples.

Children Making Patterns

Approach One: A group of first-grade children is learning to interpret a rhythmic pattern with materials. The teacher wants all the children to feel successful, and so tries to make sure they all do the task correctly. She begins a rhythmic pattern and the children join in: clap, clap, clap, shoulders, clap, clap, clap, shoulders, etc. She then leads them through the process of translating the pattern step by step. "How many times did we clap?" When the children respond with "Three," she directs them to get three tiles that are all the same color. She goes on to say, "How many times did we touch our shoulders?" "One." The teacher then tells the children to choose one tile that is a different color than the first three tiles. She continues to lead the children to make their pattern longer by using these same colors as she claps the pattern for them. If she sees that any child makes a mistake, she or another child helps them fix it.

Because the teacher set up this task to prevent errors, the activity gives the teacher very little real information about the children's understanding. All she can see is whether or not the children can follow the directions to do this particular task. In the end, she does not know what each child understands about interpreting and extending patterns.

Approach Two: The teacher asks the children to represent the rhythmic pattern in any way they wish, using whatever materials they choose. "Join in this pattern with me," she proposes. "Clap, clap, clap, shoulders, Clap, clap, clap, shoulders." After they have experienced the pattern for a bit, she says, "Now see if you can make the pattern with your math materials."

Now the teacher gains a wide range of information as the children create their various patterns. Since there is not just one way to interpret the pattern, she can see the diversity of their patterns and thus their differing levels of understanding of pattern. The teacher recognizes this as valuable information because when she knows what they really understand, she does not make assumptions and has a better idea about what they need next.

Teaching for Understanding

There is another benefit to this approach as well. When teachers ask children to try to make sense of a task, the children are actively engaged and thinking in a way that does not happen when they are being directed through a task.

This approach supports a search for meaning and understanding – merely following the teacher's directions does not.

The teacher's responsibility does not end here, however. Once she sees what the children need, she must provide additional experiences which will focus the children's attention on critical ideas. During this particular lesson, she will have those children who made a correct pattern share their work with the others. Seeing several different models will help some of the children understand the task. And over time, all the children will learn how to interpret rhythmic patterns with materials.

Let's consider another example: two contrasting approaches to teaching the concept of measurement.

Children Working with a Measuring Task

Approach One: The teacher shows the children how to measure yarn by lining up paper clips. She emphasizes that they must be careful to have each paper clip touch the one before it. As the children work, she goes around and makes sure they are all doing the task as she directed them.

When the teacher looks to see what her children are doing, she can see who followed directions, but she cannot tell who understands measurement concepts and who does not.

Approach Two: This teacher presents the task to his students by simply asking them to figure out how long the yarn is using paper clips.

Giving the directions in this way allows the teacher to see what the children think is important when they measure. The illustration shows a child who does not yet understand what it means to measure. He only knows he is to place paper clips by the yarn. The teacher would not know this if he had directed the children's actions throughout the task.

Teaching for Understanding

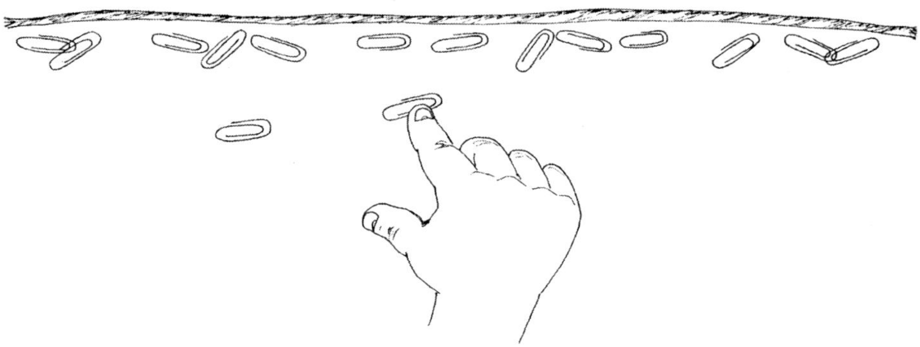

In this case, instead of just telling the child that the paper clips have to touch, the teacher interacts with them in the following ways.

"How many paper clips long did you find your yarn to be?"

"When Bobby measured the yarn he used more paper clips than you did. Why do you think that happened?"

"Did you and Bobby measure the same way?"

"What's different about the way you measured and the way Bobby measured?"

"Do you think you could fit more paper clips along your yarn?"

"Do you think you will get the same number of paper clips every time you measure?"

Teaching for Understanding

Some children will still not understand why it is necessary to have the paper-clips touch when measuring. They will not be in a state of disequilibrium if they are not yet troubled by the differing results. They will be satisfied just to accept the results as they stand. However, asking children to simply follow directions and perform the correct procedure does not deal with their lack of understanding. This only temporarily creates the illusion that they can measure correctly.

Finding out what children actually know should be viewed as a successful experience even when it provides evidence that the children don't understand. If we view children's lack of understanding and/or misconceptions as failures for ourselves or our children, we will want to prevent these failures and make sure the children do things exactly as we showed them. However, if we view a lack of understanding as part of a natural learning process that we want to be aware of, we will give children opportunities to work in ways that reveal what they truly understand.

No matter what tasks the children are working with, we can find out more about what they understand if we ask questions. The following are examples of the kinds of questions that will give insight into their thinking.

"Are you sure?"

"How do you know?"

"Can you show me?"

"Do you think that will happen every time?"

"What happened?"

"Why do you think that happened?"

Notice that these questions are not geared to getting the child to repeat the right answer. Rather, they ask a child to explain or show what they are thinking or doing. You can tell whether the questions you ask will get information about what the children are really thinking by asking yourself, "Do I already know what I want the answer to be, or am I in fact curious about what it is the child is thinking?"

We need to value the time it takes to get to know our children. Finding out what children understand is one of our most important jobs. The time invested is absolutely essential. It allows us to be much more effective as teachers as we are then able to provide the most appropriate experiences to enable "real" learning.

How Do We Deal with Children's Errors and Misconceptions?

There is probably nothing more important when teaching for understanding than the way we deal with children's errors and misconceptions. We can have the right tasks; we can provide a positive working environment; we can give children manipulatives; but we will not support children's developing understandings unless we deal appropriately with their misconceptions. The challenge is to find ways to maximize a child's learning without interfering with their sense-making process.

Children's work will have more value to them if they are allowed the time it takes to experience false starts, surprising results, and even some frustration. We must resist jumping in and disrupting the very process that can result in the most learning.

Sometimes, children make mistakes they have the ability to correct by themselves. Often simply asking a question will get them to reconsider and make the necessary changes. However, it is important that we not question children only when they are wrong. They should also have opportunities to explain what they did when they are right–so they will not assume they must be incorrect just because the teacher has asked them a question.

Teaching for Understanding

Consider the following example: The child has made a design and described it using the symbols "5 + 2 + 1."

The teacher asks a question that gets the child to look again but also allows him to explain. It is important to ask for the child's explanation. Sometimes the child is thinking about what they did in a different way than the teacher and is not actually wrong. In other situations, the children's explanations will help the teacher more clearly understand what they are having trouble with.

"Where do you see the 5?"
In this case, the child responds, "Oh, I made a mistake."
The teacher still asks the child to talk about what he did to make sure the child is not just fixing his work simply because he thinks the teacher wants him to.

"Why is that a mistake?"
The child responds, "Because there's only 4 there."

The teacher still does not tell the child how to fix the error but says, "Can you make the number sentence match the design?"

The child adds one toothpick to the design and fixes the error by himself.

Teaching for Understanding

It is not always appropriate to help children correct their mistakes. Sometimes children's errors and misconceptions are the result of natural stages of thinking and understanding that children move through as they explore concepts and ideas.

The following example helps demonstrate this point.

> Karl organized a pile of lima beans into 3 groups of ten with 6 left over. He knew immediately without counting that 3 tens and 6 were 36. When asked how many he thought there would be if he counted them by fives, he thought for a minute and said, "55." His teacher asked him to check and see. Karl moved 5 counters at a time, counted to 35 and then counted the additional 1 and said, "There's 36." He didn't seem to notice that that was the same number he had arrived at when he grouped the beans into tens and ones.

As surprising as this may be to some of us, thinking that the number changes when you count in a different way is a natural stage in children's development. Karl needs more experiences and more time to really know that the way he counts does not affect the outcome. Some teachers would think Karl's needs would be met if they asked questions like, "Did we add any? Did we take any away?" Karl knows he didn't add any or take any away. What he hasn't quite come to terms with is that counting differently doesn't change the quantity. Teaching him to say, "Counting doesn't change the number" will not enhance his understanding but instead can potentially diminish his beliefs in his own ability to figure things out.

Sometimes children's errors are the result of misunderstandings that occur because we ask them to perform tasks and procedures before they can understand them.

When Lee was asked to find out how many beans there would be if he counted them by fives, he moved one bean at a time, saying "5, 10, 15, 20 . . . ."

Lee has a misconception about what it means to count by fives. It is likely that this misconception came about because he experienced counting by fives as simply a rote process with no purposeful counting attached. Or perhaps Lee practiced counting nickels by fives, which reinforced his misconceptions. Even though Lee was able to count nickels correctly, the skill was taught prematurely. Counting nickels by fives presumes that one understands the idea that one thing can stand for many. When children count by fives before that concept is clear, it strengthens the incorrect idea in the child's mind that one can count lots of different ways and get lots of different answers! Lee's success in doing part of the task helped create the illusion that he understood what counting by fives was all about.

In order to deal appropriately with children's errors and misconceptions, we need to be able to distinguish between two kinds of knowledge Piaget described as ***social knowledge*** and ***logico-mathematical knowledge.*** Social knowledge is information that must be transmitted by telling or showing. Logico-mathematical knowledge is knowledge about relationships which must become apparent in the mind of each learner. Social knowledge can and should be presented to the children while logico-mathematical knowledge must be constructed in the minds of learners through their own reflections on their experiences.

Let's consider what kinds of knowledge children must possess if they are to solve a problem such as the following.

Sara works at the ice cream store. A family comes in and orders two single-scoop ice cream cones, four double-scoop ice cream cones, and two triple-scoop ice cream cones. How many scoops of ice cream does Sara have to serve?

Teaching for Understanding

There are various errors a child might make when working with this problem. If a child doesn't know how many scoops in a single-scoop, double-scoop or triple-scoop, they could use the wrong numbers and come up with the incorrect answer. In this case, the child is missing some social knowledge that the teacher should be responsible for providing. On the other hand, some children have trouble with this problem because they can't yet think about the various groups. They simply add up the number of cones and think that 8 is the answer. In a sense, they don't understand the question and answer a different one that they do understand. When children can't do the problem because they can't yet think of 4 double-scoops as 8 scoops, 2 triple-scoops as 6 scoops, and 2 single-scoops as 2 scoops, they do not have the necessary logico-mathematical knowledge.

When we want to determine what mathematical problem solving a child can do, we need to make sure that the child has the necessary social knowledge to solve a problem and then step back and see what the child does with that information. If we try to help children be successful by leading them through the problem step by step, we will help them get an answer, but we won't be helping them develop the ability or logico-mathematical knowledge to solve the problem for themselves.

I think one of our most challenging jobs as a teacher is deciding when to give information and when to accept the fact that the child will be unable to benefit from that information. If we consider it our job always to correct children's errors or to lead them to the right answer, we often will actually stop the children's search for meaning rather than support it.

The following examples present errors that children have made for many different reasons. The related discussions are intended to help us consider the best ways to interact with the children in these various situations.

First-grader Stephanie was making up problems using Unifix cubes and story boards representing the park. She put out 2 red cubes to represent ladybugs and 3 black cubes to represent ants. She knew without counting that she had 5 cubes all together. On her small chalkboard, she wrote "2 + 3 + 5," which she read to her teacher as "two plus three makes five."

It is not too hard to determine that Stephanie has the required logico-mathematical knowledge to do the task with meaning, but does not know how to write this down in the conventional way. Stephanie's teacher was very comfortable telling her how to write the "equals" sign, and Stephanie confidently erased the "+" and replaced it with "=".

> I once observed another group of first-grade children measuring the sunflowers they had planted. They were given meter sticks, and each was measuring and recording the growth of their plant. One of the children reported to me that his plant was 92. From my perspective, it looked about 3 inches (or about 7 cm) high. I asked the child to show me how he had measured. He took me over to his plant and showed me where he got the number he had written down in his book. The way he was measuring, the top of the plant did line up by the 92. The only problem was that he was holding the measuring stick upside down!

Let's try to figure out what information (social knowledge) and what understandings (logico-mathematical knowledge) the child lacked. Some would say the child lacks social knowledge, as he doesn't know how to hold the measuring stick right. They would teach the child what to do to prevent further errors. However, I would suggest that this error reveals the child's lack of logico-mathematical understanding. The child does not appear to understand what the numbers on the measuring stick mean. If he did, he would have been bothered by the unreasonableness of the 92.

He seemed to be doing what the teacher said to do with a measuring stick rather than seeing this as a way of recording the growth of his plant. I would rather have had the child cut a strip of paper the same length of the plant and glue that down on a piece of paper each time he measured. He could then look at the actual length of the paper strips and see how his plant was growing, rather than looking at numbers that have no meaning for him.

Teaching for Understanding

I observed a group of second-grade children measuring jars to find out the number of scoops of rice that would fit in each jar. Their teacher had taught them that it was important to be accurate and had shown them how to level off the scoop by wiping across the top of it with a pencil each time they got a scoopful.

Again, the children were more than happy to follow the directions. The only problem here was the children were not really filling the scoops full (see illustration), and when pouring the scoop into the jar, they were not concerned when a fairly large portion of the rice in the scoop missed the jar! The teacher had provided them with a procedure they did not understand, which they used without understanding or effectiveness.

Teaching for Understanding

When working with children, we need to ask ourselves, "What does the children's way of doing this task reveal about what they know and understand?" It might seem reasonable to assume that the teacher should just explain more carefully to the children what being accurate means. However, I think their way of measuring indicates they see no real need to be more accurate. The procedure of leveling the scoops doesn't mean anything to them or they would have paid more attention to that process. Children have to have a need to be accurate and some sense of what being accurate is about before a procedure intended to help them be accurate can have any meaning. One way to get the children to consider the idea that the way you fill the scoops can affect the results would be for them to compare their outcomes. The teacher could point out that Peter got seven scoops when he filled the jar and Gina got eight-and-a-half scoops. Once they have the evidence that people can get different numbers of scoops for the same jar, they can be challenged to consider why that might happen. If that is a problem for the children (that is, if it puts them in a state of disequilibrium), they will begin to pay more attention to how the jars are filled, and the procedure of leveling off will begin to have more meaning for them. They then need to continue to explore the notion of "how close is close enough" and "how can I get closer" in a variety of different situations. However, if any of the children do not have the necessary logico-mathematical knowledge to be bothered by the discrepancy, they will simply accept the idea that sometimes people get seven scoops and sometimes they get eight-and-a-half scoops.

> In another classroom, two children were working hard to figure out how many Unifix cubes long their rug area was. They were really excited to have measured and counted 428 cubes. In order to keep track of such a large number, they had broken the long train into sticks of ten. One of them came to me with several of the sticks of ten Unifix cubes to show me he knew another way to count them. "I can count these by fives, too," he said. He pointed to each stick of ten and counted, "10, 15, 20, 25 . . . ."

Teaching for Understanding

It may put you in a state of disequilibrium to think that a child can count correctly to 428, tell the number of tens he was able to make getting to 428, and still think he could also count the ten sticks by fives. But this is the kind of information we get when we listen to children. These concepts are more difficult than we sometimes realize, and children have much to sort out in their thinking about numbers. The fact they think they can switch counting techniques calls into question the level of understanding they bring to the successfully completed task of counting to 428 by tens.

This misconception shows us several things: one is that the child has not sorted out what counting by fives means; another is that he seems to think that the way you count can change the number you end up with. You may want to question the child on the spot to get him to think about what he is saying. You could ask something like, "How many cubes do you think there would be if you counted all these Unifix cubes by ones? What if you counted them by tens? How many would there be if you counted them all by fives?" Sometimes just looking at the situation will allow a child to correct himself. He might respond, "Wait, that won't work!" But more often the child will hold these conflicting ideas at the same time quite comfortably. That means he needs many more experiences over time to arrive at true understanding.

There are several things you can do to support the growth of understanding. The key is to recognize that in order to understand, one needs lots of experiences over however much time it takes. Understanding is not a one-lesson objective. It is all right to show the child what counting by fives means, but more importantly, you need to find ways to model counting by fives and other groupings in meaningful ways and in lots of different situations with many varied materials with the whole class. (Usually, one child's misconceptions are shared by others!) When you are modeling, emphasize that counting by fives means 5 at a time, counting by tens means 10 at a time, counting by twos means 2 at a time. Another way to support the growth of understanding is to have children work with smaller numbers that are easier for them to think about. While all children need opportunities to experience counting by various groupings, individual children will come to understanding at different times as they sort these ideas out in their own minds.

Teaching for Understanding

We must recognize that children's thinking and understanding evolve. We must think of concept development as happening over time. Children's ideas change and become more complete and more logical as they gain experiences and as they mature. Undeveloped ideas and misconceptions are a normal part of a child's evolving understanding. We can't prevent these misconceptions by teaching children to say words or perform procedures they don't understand. When children hold misconceptions, teachers can often help them come to new insights by providing additional experiences that allow the children to look again from another point of view. They can provide opportunities for the children to consider, notice, and draw conclusions. It is by encountering an idea in different settings and in many different ways over time that generalizations begin to form. If we insist that children must always have correct responses for concepts they are not ready to understand, they must resort to rote memory of the correct responses because they will not be able to make sense of the situations by themselves. When we try to teach children our way of thinking or our way of getting answers before they can understand, we only interfere with their sense-making process. They stop looking for their own meaning and instead look to the teacher to see if they are right or wrong.

Appropriate curriculum for young children acknowledges that they learn from their own experiences and allows them to make sense of those experiences in their own way. Mathematical competence develops in children who learn that mathematics makes sense and who learn to trust their own abilities to make sense of it. In a learning environment where children are thinking and trying to make sense of things, children can grapple with ideas they do not fully understand and can come back to problems again and again while they work to develop understanding. When giving a child a task or a question, we do not need to know ahead of time whether it will be out of reach or not. If we are willing to learn from the children's honest responses, we will be able to present ideas to them in all their complexity rather than oversimplifying them in order to ensure "success." Recognizing that children can ponder an idea before they can be expected to fully understand it allows us to challenge them in a way that we couldn't if we required them always to be right from the adult point of view. For example, we can ask children to consider which mustard jar holds the most if it doesn't worry us if they do not know. Or we can ask a child to interpret rhythmic patterns, as long as we know that they needn't be successful in order for the experience to have value for them. When the focus in on making sense of situations rather than trying to figure out what the teacher wants them to do, children will not only reveal misconceptions, but will also surprise us by their ability to figure things out if we simply allow them to do so. Considering something they don't fully understand will be a problem only if we make children feel they are not good at math, or if we make them memorize words to explain something they really don't understand.

The Teacher's Role

The teacher's job is to help children develop a full and complete understanding of concepts by providing many different but related activities. The teacher supports individual children's growth by immersing them in or surrounding them with appropriate experiences.

The activities offered children should:
- Require them to think and look for relationships.
- Have the potential to be experienced at many levels.
- Create value in being experienced over and over again.

The use of the activities should support the development of confidence and consistency as well as understanding.

Teachers need to challenge the children's thinking, helping them to:
- Make connections.
- See relationships.
- Confront their incomplete understandings.

Teachers must offer opportunities that help children:
- Focus on particular ideas.
- Make connections.
- See relationships.

However, teachers do not tell the children what to think. They structure the classroom so that what children really think and understand is revealed.

Teachers should work with small groups and individual children and engage in:

- Questioning
- Observing
- Focusing the children's attention
- Modeling or presenting information

4 Assessing Understanding

What can we learn from children's written work?

What can we learn from observing children at work?

What can we learn from student interviews?

Assessing Understanding

Some mistakes come from a lack of information. Other mistakes come from a lack of understanding.

What Can We Learn from Children's Written Work?

It has been standard practice for a very long time to determine what students know or do not know through worksheets or tests and to evaluate their progress based on the right answers they are able to produce. But what do we know about the children by looking at their written work? We can see which answers are wrong and which answers are right, but we do not have any way of knowing what the child does or does not understand about what they are doing. It is only by having them explain their work or show with models what they did that we can get a glimpse into their thinking.

As we focus on helping children develop an understanding of mathematics, we must look at more than right answers and find alternative ways to assess and keep track of their growth. We must determine what mathematics the child is using to get answers and their level of understanding of that mathematics. When we value finding out what children understand, we recognize the young child's most important work is not going to appear on paper. What young children are able to put down on paper does not always reveal what is most important about their developing understanding. Putting too much emphasis on children's production of products can take away time needed for active involvement with tasks.

If we go beyond worksheets to other kinds of written work, we can possibly gain more information. Consider the following pieces of children's work.

Assessing Understanding

The children were asked to draw a picture and write a story to go with the problem: 7 - 3 = _____.

Look closely at each of the examples. The problem is expressed very differently by each child. But what can we know about their knowledge of 7 - 3 = 4?

What most of us notice first is the sophistication of the pictures and stories. We see the long sentence and the detailed drawing of the flowers. But does this mean the child who drew flowers being hit by lightning knows more than the child who drew snakes? How do you know? If we look closer for evidence of the mathematics the students know, we see there is little difference in what we can infer from their work. Whatever differences there are in the children's levels of mathematical proficiency can only be determined by watching them at work or talking to them about what they did. Only if we actually watch them solving the problems will we know whether they knew the answer to 7 - 3 or whether they had to count to figure it out.

Assessing Understanding

Learning to write about mathematics can be a valuable process for young children, but perhaps not as valuable for gaining information about children's thinking as one might think. Young children have difficulty reflecting on their learning and usually tell what they *did* rather than what they *learned.* They might say or write something like, "I worked hard with the geoboards. I made lots of designs." They may tell about a mathematical experience. For example, "I measured the table. I used clothespins. The table was 24 clothespins long." This sounds like math, but doesn't really tell what they understood. Did they measure in a straight line? Did they line up the clothespins end to end or did they leave spaces in between? Did they count accurately? Was counting easy or challenging for them?

Because what they put in a math journal does not typically reveal their level of conceptual understanding, the journal is not necessarily a good tool for assessment. We may find the children's spelling, ways of expressing themselves, and the illustrations of interest and more indicative of what the child can do in general than simply writing answers in a workbook would tell us. However, we still must look carefully to see whether or not this tells us much about their mathematical understanding. To really determine what they know about measurement, we would need to *observe* the children *at work.*

We see then, that when young children write about math, it is often a writing experience and not a math experience. In a time period that is already too short, writing is not the best use of the time devoted to the study of mathematics. The children are still beginners at expressing their thoughts. It is a big enough challenge for most young children to begin to *talk* about their learning. *Writing* about what they are learning in mathematics, when writing itself is still a newly developing skill, can become a chore rather than an accomplishment.

As children become more mature, they will begin to move from writing about the activity they did to writing more about the process they used.

For example, when asked to determine the difference between 98 and 20, Sarah wrote about how she solved the problem.

The table sarah is 98 cubes long.

The trash is 20 cubes long.

The table is 78 longer than the trash.

10 10 10
10 10
10 10

I had 20. And I added 10s till I got to 90. then I added 8.

As children develop more writing skills they are more able to explain their thinking and put their thoughts on paper. Compared to what we can see when using worksheets where only the answer appears, the following examples reveal much more about our students levels of thinking.

All the following students solved the problem 29 + 12 accurately. But their explanations reveal they used very different kinds of mathematics to solve the problem.

Assessing Understanding

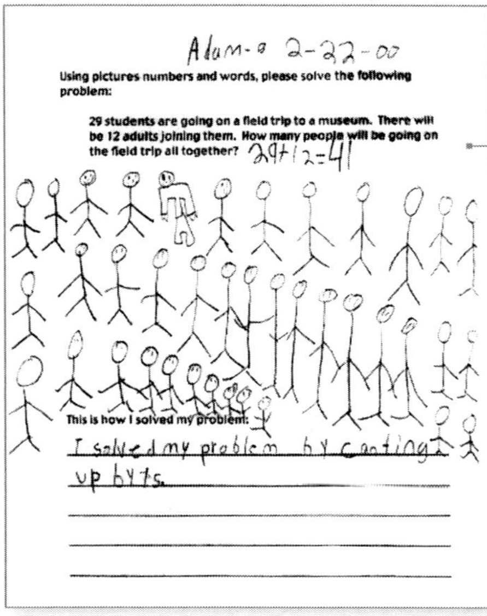

Alum-a 2-22-00

Using pictures numbers and words, please solve the following problem:

29 students are going on a field trip to a museum. There will be 12 adults joining them. How many people will be going on the field trip all together? 29+12=41

This is how I solved my problem: I solved my problem by conting up by 1s.

This child thinks of the numbers as a collection of units and needs to represent and count each item.

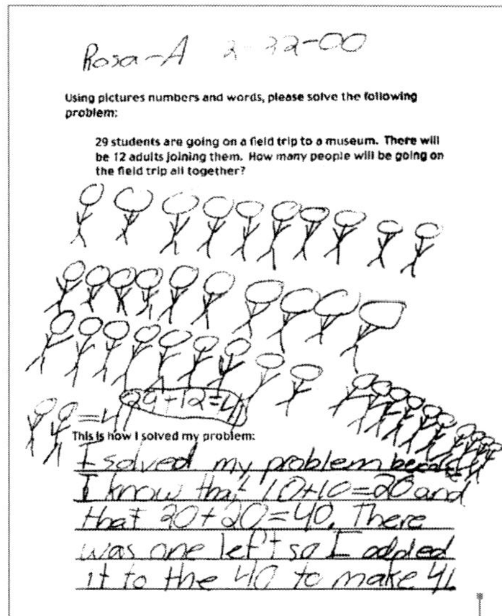

Rosa-A 2-22-00

Using pictures numbers and words, please solve the following problem:

29 students are going on a field trip to a museum. There will be 12 adults joining them. How many people will be going on the field trip all together?

29+12=41

This is how I solved my problem: I solved my problem becaus I know that 10+10=20 and that 20+20=40. There was one left so I added it to the 40 to make 41

This child has moved away from thinking of numbers as a collection of units and is able to organize numbers into groups of ten.

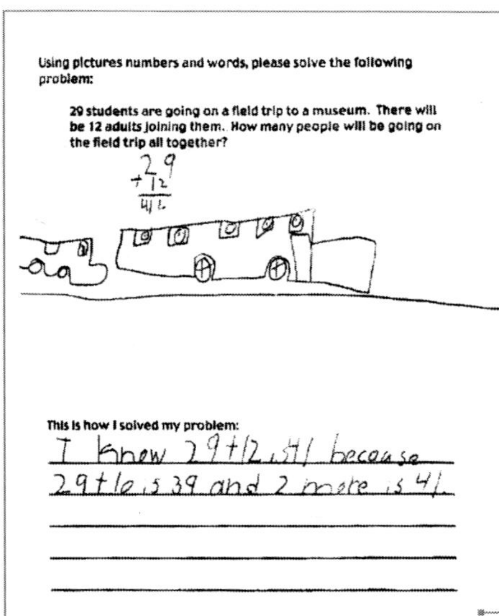

Using pictures numbers and words, please solve the following problem:

29 students are going on a field trip to a museum. There will be 12 adults joining them. How many people will be going on the field trip all together?

29 + 12 = 41

This is how I solved my problem: I know 29+12 is 41 because 29+10 is 39 and 2 more is 41.

This child recognizes that 12 can be added on easily if one breaks it up into 10 and 2. For a child that thinks about numbers in this way, this problem is almost trivial.

What Can We Learn from Observing Children at Work?

When we are interested in children's developing levels of understanding, we need much more information than a grade or a number can provide. Rather than checking to see if children have learned to perform particular procedures satisfactorily on worksheets or tests, we need to interact with them and observe them while they are working. This has value when and if we teach in ways that allow us to see what the children are thinking (as we described previously in the section, "How Can We Teach in Ways that Uncover What Children Understand?" (page 27). **We have to be more curious about what they actually know and understand than we are anxious for them to appear to have learned what we wanted them to learn.**

The key to assessing while children are at work is knowing what to look for. That means we need to be aware of the stages of children's developing understanding of concepts and their growth over time. The stages of growth for foundational concepts are predictable. In any classroom, we will see children at all stages of development for any particular concept.

Consider how children at different levels go about solving straightforward addition and subtraction problems:

> The children are using equation cards and Counting Boards. Each Counting Board represents an environment and the Unifix cubes represent children or cars or horses – whatever the children have decided their story is about. Annie places an equation card on a blue Counting Board. She pretends the Counting Board is the swimming pool at the community center. She puts the card 8 - 4 on the Counting Board. Then she places 8 cubes representing children on the blue paper. She takes 4 away all at once and writes 8 - 4 = 4 on her recording sheet.
>
> She gets another card 9 - 1 and places it on another blue Counting Board. She gets 9 cubes to represent the children and then removes 1. She writes 9 - 1 = 8. She is not doing any counting to determine the answers.

Joey is working next to Annie. He puts the card 7 – 4 on his green paper. He says his green paper is the grass at the park. Joey carefully counts out 7 cubes. After he puts them on his paper, he counts them again. Then he takes 1 cube and pretends the child is running home. Then he gets another cube and pretends it is a child running home, too. He gets another one and then another one and then another one. He counts the cubes that he took off the paper again. "One, two, three, four, five." He looks puzzled and counts again. This time when he gets to five, he takes the extra cube and puts it back on the "grass." He copies the problem 7 - 4 and then he looks at the cubes again before he writes 3.

It appears that Annie is able to solve the problems she is working at with ease. Joey needs to work hard at each step. Is Annie working at the level she should be? Is Joey? They are both able to get right answers, but there is more to know than Mrs. Smith can see at the moment she is watching them.

What Can We Learn From Student Interviews?

The *Assessing Math Concepts* student interviews go beyond finding out only whether a child can or cannot do a particular task but can identify what level of thinking the student has reached. The individual interview is the most efficient way to determine the edge of a child's understanding because you hone in on exactly what you want to know by asking the most pertinent questions to give you the most information in the shortest period of time. Once you have interviewed a child, you can then perceive so much more when you observe them doing their work.

The AMC Interviews can help teachers like Mrs. Smith interpret what she sees when observing her students at work and provide them the most appropriate instruction. She has seen that Annie can do the addition problems with ease and that Joey has to count and re-count to complete the task. But what else does she need to know to provide exactly what these students need?

Assessing Understanding

One way to find out more in just a few minutes is to give *Assessment 5: Combination Trains*. During the *Combination Train* assessment, the teacher presents a series of Unifix trains made of two colors and asks the children to tell how many of each color there is and then how many altogether. The assessment is designed to help her find out what combinations the students know without counting for Numbers to 6, to 10 and to 20. She will also be able to see if they can use what they know about one combination to figure out a related combination.

Mrs. Smith calls Annie over to her while the other children are working at various math stations and in just a few minutes is able to gain important information about Annie's understanding and proficiency with combining numbers.

The first Unifix cube train is made of 2 blue, 2 yellow, 2 blue, and 3 yellow cubes. 

Mrs. Smith shows the first train to Annie.
Mrs. Smith: How many blue cubes?
Annie: (instantly) 4.
Mrs. Smith: How many yellow cubes?
Annie: (instantly) 5.
Mrs. Smith: So, there are 4 and 5, how many is that all together?
Annie: (instantly) 9.
Mrs. Smith: How did you think about that?
Annie: That's easy. 5 and 5 is 10, so 4 and 5 is 9.

Assessing Understanding

Mrs. Smith shows the second train to Annie.
This train has 3 blue, 3 yellow, 3 blue, and 4 yellow cubes.

Mrs. Smith: How many blue cubes?
Annie: (instantly) 6.
Mrs. Smith: How many yellow cubes?
Annie: (instantly) 7.
Mrs. Smith: How did you think about that?
Annie: Well, 3 and 3 is 6 but it is 3 and 4 so that's 7.
Mrs. Smith: So, there are 6 and 7, how many is that all together?
Annie: (instantly) 13.
Mrs. Smith: How did you think about that?
Annie: I know 6 and 6 is 12, so 1 more makes 13.

Mrs. Smith shows the third train to Annie.
This train has 4 blue, 4 yellow, 4 blue, and 4 yellow cubes.

Mrs. Smith: How many blue cubes?
Annie: 8
Mrs. Smith: How many yellow cubes?
Annie: Oh... 8 again.
Mrs. Smith: So, there are 8 and 8, how many is that all together?
Annie: 8 and 8 is 16.
Mrs. Smith removes the Unifix train
Mrs. Smith: How many there would be if we had 8 blue and 9 yellow?
Annie: That's 1 more so it's 17.

Assessing Understanding

Annie has no difficulty at all with the assessment. She knows the combinations to 20 and is able to use what she knows about doubles to solve related problems. Mrs. Smith now sees that Annie needs to work with problems with bigger numbers. She will go on to *Assessment 6: The Hiding Assessment* to see if Annie can decompose numbers as easily as she combines them. Depending on how Annie does, she may decide to assess her using the *Ten Frames* assessment to see if she still needs to work with making tens or is ready for 2-digit problems.

When Mrs. Smith assesses Joey, she finds he counts for every combination – even 2 + 2. She decides to follow up with the *Counting* assessment to see if he needs more work with counting rather than with working with equations. He would still be able to work with the Counting Boards alongside his friends, but he could use numeral cards instead of equation cards.

Mrs. Smith assesses the rest of her class and finds that some of her students need to work with sums to 6, some need to work with sums to 10 and a few need to work with sums to 20. No one is quite at the level that Annie is working, although a couple of students are working with sums to 20 and will be able to move on soon.

The individual interviews help make the teacher observations more effective. Mrs. Smith now has information that helps her tune into each child as they work. She knows that Peter has been working with sums to 6 but is doing well and is about ready to move on to sums of 10. She can see that Annie is challenged by making tens and is working to solve them without needing to actually move the cubes. Joey is developing confidence counting so when she decides to give him equation cards, he will be able to work with them without needing to count and recount all the time.

In order to accommodate all her students at the appropriate level, she uses color-coded equation cards. All of the children can still choose the Counting Boards, but some will use green cards with sums to 6, some will work with yellow cards with sums to 10 and some will use the pink cards with sums to 20. Joey and two other children will be choosing numeral cards for a while before working with equation cards.

5 Classroom Instructional Settings

Which instructional settings should we use?

Whole Group Time:
What kinds of work should be done with the whole class?

Small Group Time:
How can we work effectively with small groups?

Independent Work Time:
Why is it so important?

Classroom Instructional Settings

Make sure the children are continuously thinking and engaged.

Which Instructional Settings Should We Use?

There are a variety of classroom settings that can be used to support children's learning of math concepts. Some activities are best presented in a whole group, teacher-directed setting; other activities should be offered as independent work; and some activities should be presented by the teacher to small groups of children. Each of these settings serves different purposes and provides different opportunities for the children. I think of the various settings as options I can use depending on what I think will allow me to make the best use of our time.

Working with the whole class at once is the organizational structure that is effective for giving certain kinds of instructions and directions that you don't want to have to repeat over and over again. There are also ongoing activities, such as estimating and opening activities that can be done effectively with the whole class. The independent work time can be the most effective and most productive setting, as it is the organizational structure that allows us to most closely match children's individual needs, styles, and interests. Working with a small group allows you to really watch, interact with, and respond to individual children during a "teaching" time.

Whole Group Time: What Kinds of Work Should Be Done with the Whole Class?

There are certain activities that are effectively and efficiently done with the whole class. However, we must be careful to make sure that the children are continuously thinking and engaged during this time.

Be aware of how you work with the whole group, especially if you find that the whole class is responding as one voice. A group response may be appropriate, but it usually indicates a memorized response, one that individual children may or may not understand. These kinds of answers may indicate knowledge that could be of value, but they are not responses that require thinking and understanding. One of the biggest problems with whole group work is the assumption we often make that all (or even most) of the children understand what they are saying in chorus. For example, when you ask a group of children, "How many tens and how many ones?" when they are looking at a number like 76, you will most likely get correct responses. It will only be in other settings like independent work time that you will be able to tell if these words have any meaning for them.

The following are the kinds of activities that are appropriate in a whole class setting.

Introduction or Review of Activities

(Even the youngest children can learn how to do tasks this way.)
It is an efficient use of time to introduce activities to everyone at once. If there are individual children who don't learn how to do the task when it is introduced to the whole class, the other children will be able to help them. Make sure the children know they should ask another child if they are unsure of the directions, rather than coming to you. That way they will learn to be independent, and you will be able to work with children who really need your help.

Ongoing Activities

There are certain ongoing activities that work well as whole group activities, such as opening activities, estimating activities, and graphing. However, because children need to be involved in doing their own work as much as possible, keep this time short. Remember to get the children thinking as much as possible rather than simply responding along with the group. For example, all the children can be involved in estimating the number of walnuts it takes to fill a jar. Even though the teacher is the one to fill the jar while the children count the walnuts, there is a sense of active participation as they watch the jar being filled and compare what they thought was going to happen with what actually happens.

Classroom Instructional Settings

Shared Experiences

There are some activities that the whole class can experience at once if there are enough materials for all the children to work with at one time. For the tasks to be appropriate for the whole group, all the children should be able to do the same task, each in their own way. They should not be kept together or directed step-by-step through the tasks. For example, the children can work with activities such as sorting collections or making increasing patterns with a variety of materials.

There is a type of shared experience I call mathematical events. I define a mathematical "event" as any experience that involves children in doing mathematics, but is not a part of any set of experiences that supports the development of a particular concept over time. The teacher chooses the task because it looks interesting and believes her children will enjoy the challenge. The "events" often involve problem solving. While the term "problem solving" can be interpreted in a multitude of ways, I am using it to refer to any question that engages a child in a task to find something out. These special "events" can be interspersed throughout the year for a day or two at a time. Children's literature is often a source for these tasks. (See *Posing Open Ended Problems for Primary Students,* listed in the References, for engaging and appropriate tasks for your students.)

Problem solving and other mathematical events can be an important part of children's experiences with mathematics. They provide them with opportunities to see themselves as "mathematicians" and to use the mathematics they have learned. We can employ these experiences to help us assess children's needs, as their method of work often reveals which concepts and skills they have sufficiently internalized to be useful as tools and those they are not yet able to use.

Again, I have some words of caution. We must take care to ensure that the use of children's literature and various investigations enhance the learning of mathematics rather than simply distract from the important ideas we want children to confront. We must be careful in selecting these experiences to make sure that the mathematical understanding gained is worth the time involved in doing the task. These experiences should not replace ongoing work with concept development.

Small Group Time: How Can We Work Effectively with Small Groups?

Working with a small group allows you to really watch, interact with, and respond to individual children during a "teaching" time. You can use this instructional time in three ways:

- to present activities that are difficult to introduce to the whole group
- to observe more closely and thus to assess the needs of children without having to do individual interviews
- to provide experiences to meet the needs of a particular group of children

Small group instructional time is particularly important for some children who find it more difficult to be as engaged and focused when they are part of a large group than when in a small group.

Try to be flexible in the use of small group time. Groups do not need to be set up permanently but can be formed when particular needs arise. You can meet for very short periods with some groups and for longer periods with others. You may want to meet with selected groups for several days in a row if they need extra help or an additional challenge. Teachers have told me that they don't think it is fair to spend more time with some children than with others. They are concerned that they are always responding to the ones who are struggling and trying to find ways to challenge those children who are way ahead of the others. They worry about the students in the middle who are always getting overlooked. I agree that it is important to meet the needs of every child in our class. However, that should not be measured in time spent with the teacher. Rather, our goal should be to find a way for all children to work at the edge of their understanding.

Some groups will need to meet with you briefly every few days and will be able to accomplish much by working independently. Other children will need more teacher guidance to focus on the concepts and will need to meet with you more often.

Prepare the Children for Work with Small Groups

Before you can work effectively with small groups, you will need to prepare the class. The following guidelines will help you teach the children how they are to handle themselves during this time.

- **Make sure the children know how to work independently before you expect them to work without bothering you when you are busy with a small group. Have them practice not interrupting you ahead of time.**
  Before you actually pull out a group, tell the children that you are planning to work with small groups during math time. Let them know it is important that they not interrupt you when you are working with the group. Tell them you are going to watch them that day and see if they can work independently without bothering the teacher. Spend at least one period having them show you how well they can work. If any problems come up, discuss with them how they should be handled if you are busy. The first time you work with a small group, make it a very short period so they can be successful.

- **Make sure the children know all the independent activities before you pull out a small group. Don't pull a group out if you have just introduced new activities.**
  The activities need to be familiar so the children can be expected to know exactly what to do without bothering you.

- **Keep the small group time short (about 15 minutes).**
  If you keep the group time brief, you can more easily expect the other children to be successful.

- **Don't label the children by having permanent low, middle, and high groups.**
  Sometimes your small group time can include children with varying needs, and other times you will want children with similar needs to be together. For example, if you wanted to gather a few together so you could more easily observe how they did a particular task, or if you wanted to teach an activity to a small group which they would then do independently, you would require a heterogeneous group.

If you wanted to work specifically with children having trouble with one-to-one correspondence or children who needed to work with number combinations of six, you would call the children with those particular needs together.

- **Work with small groups only when you have a specific need, rather than having a permanent schedule.**
  It is important to think of the various classroom organizational structures as tools to use for different purposes when we need them. This allows more flexibility than setting up a permanent schedule that dictates what we do and how we do it. For example, if I were to form four permanent groups to work with me one day a week, I would be focused on what I was going to do with each group for their block of time rather than on how I might meet the needs of my students as they became apparent to me. If I form small groups only when the need arises, I would be able to work with a small group of children who were having trouble with a concept for three days while the rest worked at independent activities. Another time, there might be a group of children ready for a challenge that could meet with me only once and be ready to work on that challenge independently. At other times, I might want the whole class working together for a few days on something, and I wouldn't want small group sessions at all for a time.

Establish a Routine for Working with Small Groups

On those days when you plan to work with a small group, begin the period by gathering the whole class together on the rug. Remind the class of what they will be working on at the independent stations and tell them you will be working with a small group and should not be interrupted.

Ask the group of children you want to work with to stay on the rug. Give them a task to do while they are waiting for the lesson to begin. I often just let them work with the chalkboards, writing or drawing whatever they want. Other times I might give them a particular job like, "Count and write the numbers starting with 25," or "Choose a numeral you need to practice writing," or "How many ways can you write equations that add up to 7?" Excuse the rest of the children a few at a time to choose where they are going to work.

Watch the class for a few minutes to make sure they have the materials they need, know what they are supposed to do, and are settled in.

Work with the small group for 15 minutes or so. Then dismiss them to choose an independent station for the rest of the period or, have them stay where they are and practice a task you just taught. Go around the room and interact with the children who have been working independently. Make note of any needs you need to handle with either the whole class, in a small group time, or individually. When work time is over, ask the children to clean up and gather back on the rug. Spend a minute or two reviewing how the math time went, pointing out what went well that allowed you to work with the small group, or discussing any problems that arose. This routine works even without other adults in the room. However, if you do have an aide or parent volunteer, they could monitor the individual stations, asking children questions while you work with the small groups.

A Plan for Working with Small Groups

1. **Gather the class together.**
   Begin the period by gathering the whole class together on the rug. Remind the class of what they will be working on at the independent stations, and that you will be working with a small group and should not be interrupted.

2. **Ask the group of children you want to work with to stay on the rug.**
   Give them a task to do while they are waiting for the lesson to begin. I often just let them work with the whiteboards, writing or drawing whatever they want. At other times, I might give them a particular job, such as "Count and write the numbers starting with 25," "Choose a numeral you need to practice writing," or "How many ways can you write equations that add up to 7?"

3. **Excuse the rest of the children a few at a time to choose where they are going to work.**
   Watch the class for a few minutes to make sure they have the materials they need, know what they are supposed to do, and are settled in.

4. **Work with the small group for 15 minutes or so.**

5. **Dismiss the small group.**
   Have the children choose an independent station for the rest of the period, or have them stay where they are and practice a task you just taught.

6. **Go around the room and interact with the children who have been working independently.**
   Make note of any needs you will have to handle either with the whole class, in a small group time, or individually.

7. **When work time is over, ask the children to clean up and gather back on the rug.**

8. **Review how math time went.**
   Spend a minute or two reviewing how the math time functioned, pointing out what went well that allowed you to work with the small group, or discussing any problems that arose.

Independent Work Time: Why Is It So Important?

Independent Work Time is at the heart of the math program. It is the setting where children can benefit from spending the most time. It is in this setting the teacher can make sure the children get the experiences they need at their own level, and the teacher has the opportunity to interact with the children at their level, providing support and questions most appropriate for each.

I think of the math program as moving from one set of related activities focused on a concept to another set of activities focused on another concept. So, for example, a class may be working with beginning addition and subtraction for several weeks or months followed by work with numbers to 20. The set of tasks is expandable, thereby allowing children to work at the right level and move on as they can. It is by surrounding the children with multiple ways to experience a concept that they develop the insight and proficiency they need.

In fact, the whole group activities are primarily designed to provide children with the information they need in order to work well within the independent stations. Sometimes the function of the whole group work time is to focus them on what is important about the work they are currently doing and sometimes it is to give them the experiences they will need to work with the upcoming concept. The teacher-directed work with small groups of children will provide support to those who either need more work with concepts the class has worked on previously or to help them work more productively with the current activities. Occasionally it can provide a challenge to children that they can apply when working independently, or sometimes it serves to gather information about their needs that you have not been able to identify by observing them at work.

Children need numerous experiences with a set of related activities that give them the opportunity to gain a broad and complete understanding of whatever concept(s) we want them to learn. And they need lots of time to work with these concepts so they can develop understanding and facility. Children will not have the opportunity to develop a deep understanding of math concepts if they are given a random collection of experiences in which no concept is fully developed and individual needs are not considered.

The time children spend working independently is very important. This is when they are fully responsible for the work they are doing, and it is when they get the practice and experiences they need. Most of the time, they should be working with tasks they have chosen from a set of related tasks.

For independent time to be as productive as possible, children need to know how to work independently and to feel responsible for their own work and learning. What I have referred to in my work with teachers as Self-Directed Exploration is called Self-Regulation by some researchers. The impact of the student's ability to regulate one's self goes far beyond mathematics and is supported by research into what makes students successful in school.

6 Developing Self-Directed Learners

How do we help the children learn to work independently?

How do we transition from exploration to concept development?

Developing Self-Directed Learners

I feel more relaxed about the time allotted to self-directed exploration if, at the same time, I begin concept development work through whole group teacher-directed activities for 5 to 15 minutes a day.

How Do We Help Children Learn to Work Independently?

If you are going to ask children to be self-regulated learners, to work responsibly at independent work time and are going to use materials to teach mathematics, then it is important that you start every year with a period of several weeks when children are allowed to work with the math materials using their own ideas. We refer to this time as self-directed exploration.

Begin with Self-Directed Exploration

Before children can focus on particular tasks the teacher has in mind for them, they must be given the opportunity to explore materials in their own way. Teachers everywhere have learned for themselves how important this time is. While it may look like play time to those who do not know the purpose or benefits of self-directed exploration, this can be one of the most valuable parts of the math program.

There is much for the children to learn. Not only do they become familiar with the various materials, but they explore mathematical ideas and learn what the expectations of the classroom are. And, they learn to be in charge of and engaged in their own learning. This time is critical to establishing the learning environment and will impact what happens the rest of the year.

Make Note of the Mathematical Ideas that Children Explore and Discover

When we watch children working during self-directed exploration, we will see examples of their creativity and problem-solving abilities. But beyond that, we need to look for the evidence of the mathematics that is naturally occurring. As children become familiar with the attributes and possibilities of the materials, they will naturally begin to:

- Create pictures and designs
- Sort in a variety of ways
- Measure
- Build increasingly complex structures
- Encounter problems of balance and symmetry
- Create and extend patterns
- Count and compare quantities
- Use phrases like "longer," "shorter," "not as much as," "more than," "just the same as"

There is a certain amount of trust required from the teacher in order to feel confident that the children are learning in this kind of a setting. Somehow we know, when we give children blank paper and pencil or paints, that the creative writing or paintings they produce are more valuable than following someone else's ideas would be. We would not be as impressed with work generated by "fill in the blanks" or a "paint by numbers" product. We respect winners at the high school or college level who participate in problem-solving contests that require them to do such things as build models of bridges or other kinds of structures. Young children deserve the same kind of respect for the work they do with materials. By working in this way, children discover the potential of the materials and the power of their own ideas.

Help Children Develop a Work Ethic and Sense of Responsibility

Self-directed exploration time is important not just because of the mathematics to be learned, but also because it provides a wonderful opportunity for children to become familiar with the culture and expectations of the classroom. During this time they will be learning the importance of many activities:

- Self-direction and self-discipline
- Decision-making
- Choosing tasks and staying with them
- Working independently
- Working hard, staying engaged (without the teacher's attention)
- Working with others–sharing materials and space
- Setting up and putting away
- Inventing, creating, and problem solving

Establish the Following Procedures

You must take certain steps and establish certain procedures to create this working climate.

- **Start with just a few materials.**
  Introduce two or three different materials the first day and then present the rest of them gradually over a period of several days.

- **Let the children know your expectations for working with the materials.**
  It is very natural for children to work creatively with materials. They know what to do even without direction from the teacher. However, the tone you set will have a big influence on the attitude they bring to the task. Emphasize the importance of working hard. Since I don't want to imply that this is play time or a time for children to do anything they want, I often use the phrase, ". . . working hard using your own ideas," rather than saying, "You can do what you want with the materials."

There are a few rules that the children need to understand if this time is to go smoothly. The instructions for using the materials need to make sense to the children. They need to see that the guidelines help everyone get along better with each other. I share the following basic rules with the children.

- **We do not throw things in the classroom.**
  Our classroom needs to be a safe place.

- **We share the materials. We don't need to hoard.**
  We take only what we need as we need it.

- **We clean up before moving to a new station.**
  This leaves the station ready for the other children so they can have a turn using the materials.

- **We mess up our own work only.**
  When people create something, they usually want to take it apart themselves. When someone else messes up our work, it feels like they are wrecking it.

- **Have the children set up.**
  We want children to take as much responsibility as possible for their work and actions. One of the ways we encourage this is by giving children responsibility for putting out the materials at the math stations.

When it is time to set up the math stations, have several students put them out at the designated places while you continue an activity (such as rhythmic clapping or counting) with the rest of the children. Some teachers have a group of three or four children assigned as helpers for the week. Other teachers have certain children be helpers on Mondays, a different group on Tuesdays, etc. Still other teachers choose children as they need them. They select children depending on who has been working hard or who needs the opportunity to help.

- **Let the children choose where to work.**
  After the student helpers have set up the materials, dismiss a few children at a time to choose the materials they want to work with. (There should be enough places to work so that even the last child still has a choice.) Allowing the children to make choices is critical in establishing the environment we wish to create. When children choose where they work, they are more interested, involved, and committed than if they feel they are just doing what the teacher asked them to do. Because they chose where to work, we can expect them to be interested and to work hard.

  It may seem more efficient to assign children to the materials so they will become familiar with all of them. However, asking children to be creative with a material they are not yet interested in proves counterproductive in the long run, and does not move us forward in developing the independence that will prepare them for later self-directed work.

  Teachers have found that children will naturally explore all the materials over time. Those who are interested in a material find all kinds of ways to explore it, and other children then become intrigued. One way to ensure interest in a material that has not yet been discovered is for you to work with it. Many of the children will love to work side by side with you. There may also be times when you ask the whole class or a small group to work with and discuss a particular material. But that is a distinct experience from this work time, where enabling the children's choice is essential.

- **Allow the children to choose to work alone or with others.**
  Children generally decide for themselves whether to work alone or with others. Allow children to work together if they are able to make good

use of their time. I do not assign children to work together, because it is often beyond their capacity to sort out the problems that can arise when two children want to do different things. It is rare that a child always wants to work alone. They usually will move naturally from one mode of working to another.

- **Expect the children to be accountable for working hard.**
  Because one of our goals is helping children learn to make responsible choices, certain parameters should be observed. The children need to understand that making choices is not a license to do anything or choose anything. They need to know the range of the choices: in this case, the choices are the math materials, not Legos or Tinker Toys. There may be another time in the day when all the materials in the room are available to the children, but not during math time. One reason for this is that our long-range goal is for the children to be ready to do specific activities with the math materials. We want them to be familiar with these particular materials so we can use them for concept development.

- **Allow the children to move from place to place whenever they are ready to work with something else.**
  Children's levels of interest and ways of working vary. Some will choose the same material for several days in a row; others will want to work a little with many different materials before they can really get engaged with any particular one. It usually takes a few days for all of the children to get totally involved – but it does happen.

  If you have one or two children who can't seem to get involved, you will need to intervene. There are several different things that you can do to help these children. Some may be overwhelmed with the number of choices. Give them two or three things to choose from. Others may feel they don't know how to do it "right." Sit down and work with these children for a time. If they are having trouble getting started, you might make a simple structure and ask them if they want to copy it. Then have them make something for you to copy. Help them see there is no right or wrong way for them to work with the materials.

It is usually necessary to limit the number of students allowed to work at each station, so children will not run out of materials or space. I usually limit the station to six children at a time. Sometimes the number of chairs around a table will indicate six places. When there are no chairs, the children count to see if there is room for them to work at a particular station.

- **Teach the children a signal to respond to when you need their attention.** Whenever you have noise and activity, you need a way to get the children's attention. You will probably want to teach the children to respond to some kind of signal. Ringing a "magic" bell is one way to get children's attention. Or you may want to quietly give a direction such as "hands on your shoulders" or "touch your nose." Children catch on quickly once they see others responding. One of my favorites is an echoing response. The teacher claps a rhythm (clap, clap—clap, clap, clap) and the children respond by clapping the same rhythmic pattern.

- **Tell the children they don't need to come and get you.** One of the major goals for self-directed exploration is for children to develop the ability to work independently. This will allow you to work with individual children and small groups as needed. In addition, we want the children to learn not merely to perform for you, but to be engaged learners in tasks that absorb their attention. This attitude is actually very important for the children's future. Think about those employers who advertise for "self-starters."

  You want the children to know you are interested in what they are doing, but not to the point that they seek out your attention all the time. Students need to learn that it isn't necessary for you to notice and comment on everything they do. I recommend that you not make yourself available to them whenever they want you. Instead, tell the children that you will come by and see what they are doing when you are ready do to that.If they come to get you, reassure them that you will be coming to see everyone's work in a few minutes. Soon they will trust that you will come by even if they haven't asked you to come and see their work. Your attention will become less and less important to them as they become more and more focused on their work and the sense of accomplishment it brings them.

Developing Self-Directed Learners

Developing Self-Directed Learners

Give the Children Plenty of Time to Work

It is important that you allow at least 45-50 minutes for this "math work time." You will get a higher level of involvement from the children if they don't feel it will be time to clean up before they really get started. If you teach kindergarten for a half-day, you may feel you do not have enough time in your day to devote this much to mathematics. Some kindergarten teachers have found it is better to have math for longer periods of time three days a week than for a short daily period. If possible, arrange your schedule for some flexibility. When children are really engaged, you can allow them more time. If they are having trouble focusing on the tasks, be ready to clean up early. However, be patient. There is often a period of restlessness right before a renewed level of involvement. If you feel your children are not being very creative, don't give up too soon. Join in with them and enjoy the materials yourself. It may take several weeks before children really discover the endless possibilities of the materials and their own abilities to create.

Get to Know Your Children

At first you may not know what you should be doing when the children are busy with their own ideas. Your most important job is to stand back and observe. This is a wonderful time for you to learn about your students. The information you gain will help you immensely in making decisions about your children and their needs.

Watch to see how they work. Do they work alone or with others? How do they respond if another child accidentally bumps their work? Do they stay focused on one material, or do they move from material to material? Do they share easily, or do they need to be reminded? Do they like to talk about what they are doing, or do they work without talking? Are their creations simple or complex? Do they like to organize or build with symmetry, or are they more random in their approach? Do they make designs or structures, or do they think of what they build as houses or castles or robots?

You may want to take notes on what you observe about your children. Some children are actually motivated by seeing that their work is important enough for you to write down. Your presence and interest in their work will help communicate to the children that what they are doing is important. However, because we want children to work for the satisfaction of their own learning and not just to please the teacher, do not judge or comment on their work other than to say something like "It looks like you've been working hard," or "Was that fun to do?" or "Tell me about what you are working on."

Follow the Children's Lead When Interacting with Them

Resist the urge to become over-involved in the children's work. When you do engage a child in a discussion, try to elicit conversation from them about what they are doing, rather than just imposing your thoughts on them. Sometimes simply asking, "Tell me about what you made (or are working on)" will be sufficient and then the child will either respond eagerly with some detail or very briefly, getting right back to work. When the child is at a point in the work that they want to talk about it, finding the words to tell you what they have done or has made can be beneficial and also can provide an opportunity for you to model the use of certain words in the context of what the child is doing. For example, if a child says, "I like these tiles. They are really shiny and soft," you could comment, "I don't think these tiles are soft like a teddy bear, but they are really smooooth. When I feel them, there are no bumps. Is that what you were noticing?"

Sometimes our questions interrupt rather than enhance or support the children's work. I had a friend who once said, "I drop questions on my kids. Some of them stick and some of them slide off." I think this is a good attitude with which to approach your interactions when the children are working. Sometimes you will stimulate a new idea or exploration by asking a question; other times you won't. Examples of open-ended questions you can ask are:

"What do you think would happen if . . . ?"
"How many do you think it would take to . . . ?"
"Do you have another idea . . . ?"

Intervene if Necessary

Because this is such a natural way for children to work, you will find most are actively engaged during this time. Discipline problems occur less often than they might in other situations. While the children should be allowed to work with minimum interruptions, they should not be allowed to be unproductive or disruptive to others. The children who are not handling this period of time appropriately will stand out, and you can intervene as necessary. If children sometimes wander or fail to become engaged, ask them to tell you where they are planning to work. You may need to walk with them to the place where they will work, check on them in a few minutes to see how they are doing, and in some cases take away the privilege of moving freely from station to station. Any children who choose to work together but distract each other and do not get to work may have to have the privilege of working together taken away for a period of time. Later you can provide opportunities for them to try again to work productively together.

Teachers who envision self-directed exploration as a version of playground activity will be pleasantly surprised at the hard work and productivity that most of the children will display. As long as they know this is not play time and know what the expectations are, you can expect this to be a busy, active and yet rewarding time for everyone.

Establish Guidelines for Clean-Up Time

The children should be responsible for clean-up time and for putting materials back where they belong. For this to go smoothly, it is necessary to have some guidelines. Children need to know that they are to clean up their own creations and that it is not helpful to have someone else come and take apart what they have worked hard to create. The children must know where to put the materials, and where they are to go after they have cleaned up their work area.

It helps to have something interesting going on, such as a song, rhythmic pattern or finger play so they will be motivated to clean up quickly and go to their places on the rug. Sometimes two or three "housekeepers" can go around and make sure all the pieces are picked up and all the tubs and baggies of material are back where they belong.

Discuss How Things Went

At the end of the math period, it is very important that you gather the children together to discuss how things went. While usually a very brief time, this exchange is vitally important. It is another opportunity for you to let the children know that you value behaviors like hard work, sharing, getting along, and using their own ideas. You can tell them what you noticed that showed you they were using this time well. A few children can share what they did or how they worked. Even though just a few children share each day, this sharing validates the importance of the work time to all the children.

If there happens to be a problem on a particular day with noise or too much wandering or silliness, the children will usually be able to identify the problem and discuss what needs to happen in the future. When things go well, they will be able to recognize that as well.

Resist allowing this time to become a sharing time for everyone. It is difficult for children to listen to each other for long periods, and this is not the best use of math time. It is more meaningful for young children to talk about their work while they are actually working. The children should have many other opportunities to talk about their work and to learn to express themselves as they interact with others or with you during their work time.

First-grade teacher Kathy McGrath adapted the idea of "working levels" to meet the needs of her first-graders. A few weeks after school starts, when the children have had enough experiences to be able to be reflective about the math period, she introduces them to Working Levels. Working Levels are designed to help raise the children's awareness of their behavior, to make the expectations more explicit, and to give focus to discussions about what it means to work hard.

There are five Working Levels that the children become familiar with:

Level 0	Level 1	Level 2	Level 3	Level 4
Bothers others	Not working	Works when reminded	Responsible Respectful	Responsible Respectful Challenges self

If you decide to use the Working Levels, you will find that as the children learn what it means to work hard, the discussions that are generated are more important than the levels themselves. Be specific and concrete with the children as you discuss such questions as "What does it look like to work hard?" "What does it look like to be responsible?" "How can we be helpful to others?" "What kinds of things would not be helpful?" "What will we see?" What will we hear?"

Together with the children, you will identify such things as: "Working hard means staying with a job and not wandering around." "It means concentrating on what you are doing and not just talking." "It is not helping others if you mess up their work or do their work for them." "It may be helping others if you helped them pick up materials they spilled, but you need to ask first and make sure it is all right to do that."

The children should decide for themselves how they think they did and share their feelings about their work only if they want to. Young children are usually quite honest and are willing to acknowledge if they had difficulty settling down to work on a particular day or if they feel they worked especially hard.

The levels are not intended to be used to label the children themselves but to give them a way of thinking about how that particular day went and to help them become more aware of their behavior. This is not intended to put pressure on children to be perfect, but to help them be honestly reflective.

Self-Directed Exploration

1. **Decide what materials you are going to introduce.**
   You may want to include books, chalkboards, paper, and markers at first and gradually replace them with math materials.

2. **Emphasize to the children that this is a work time.**
   Discuss what hard work looks like. Make your expectations clear.
   - Share the materials.
   - No throwing.
   - Don't mess up other people's work.
   - Clean up your work.

3. **Support the growth of independence and self-direction by reminding the children not to come and get you.**
   Reassure them that you will see everyone's work as you walk around.

4. **Observe the children at work.**
   Redirect any children that need reminding of appropriate behavior.

5. **Have something for the children to do as they finish cleaning up.**

6. **At the end of the period, gather together and briefly discuss how it went.**

7. **Expect the children to become more involved over time.**

How Do We Transition from Exploration to Concept Development?

Give the children opportunities to work independently with the materials several times a week for several weeks. Kindergarten children will need at least 12 weeks, first-grade children will need six to eight weeks, and second-grade children will need four to six weeks. The level of involvement and the complexity of the children's work evolve over time. The tone and expectations you set during self-directed time will establish the classroom environment for the other activities you plan to do for the rest of the year.

Don't rush through this. It is important to establish a productive work environment. What you do now will affect the whole year!

Although you may be anxious to move on to the more structured work with concept development, if you leave exploration too soon, you will not get all the benefits from it. It is important to stay with this long enough to see that the children have learned to work hard and independently.

Sometimes teachers attempt to move the children too quickly to concept development activities. When that happens, the children let us know. They lack the focus and enthusiasm for the work that we might expect. Even very experienced teachers have to be reminded of this on occasion. A friend of mine taught a first- and second-grade multi-age class for the first time after years of teaching first grade. After three weeks of self-directed exploration, she was feeling the need to begin pattern activities with her second-grade students sooner than she would have moved her first-graders in the past. She worried that the second-graders were ready for more of a challenge and she didn't want to hold any of them back. However, she found that the children had difficulty focusing on the tasks they had been assigned. They didn't seem deliberately resistant, but they didn't get much accomplished. She picked up on these clues and gave the children two more weeks of exploration time. This was time well spent as it was followed by an exciting period when the children worked eagerly with increasing patterns and patterns on the 00-99 chart.

Developing Self-Directed Learners

If you find out the hard way that you have moved the children to structured activities too soon, go back to self-directed exploration for a time. Then, when you reintroduce the structured activities, you will see the difference!

When the children have shown that they can work independently in pursuing their own ideas, you can begin making the transition to structured activities by providing opportunities for directed exploration at some of the stations. For example, you could put out recording materials with the pattern blocks or geoboards so the children can make records of their designs. You will get an idea of which children are ready for more structure by the level of interest at these stations.

When you have some stations where children are allowed to use their own ideas and other stations where they have a specific task to do, make sure the children understand the difference between them. It is not appropriate for you to allow them to work freely at a station if it is set up for a specific task. You want them to understand that sometimes there is particular work that you expect them to do.

You can have both free exploration and structured tasks going on at the same time if it is clear when they can choose what they do with a material and when the material is there for a specific task. For example, while the pattern blocks on the floor may still be available for exploration, the pattern blocks that are on the table along with a tub of recording materials are used to make records.

Gradually replace all of the self-directed exploration stations with specific tasks by introducing a few at a time until all of the stations are concept development activities. You can put the materials you need for a particular station in a tub or baggie and have the children set up the stations. Establish particular places for the stations to be delivered. This helps the children know what the choices are and keeps them from wandering around the room trying to decide where to work.

Developing Self-Directed Learners

Periodically throughout the year, give children some time to work again with the materials using their own ideas. During this time, you will often see evidence that they are applying the concept development work they have been doing. Some teachers also allow the children to work with the math materials when they have free time during the day. That way, exploration can continue concurrently with concept development.

A Math Time Routine

1. **Gather your whole class together on the rug.**
2. **Do a short activity (5-10 minutes) together.**
   This should include a daily Number Talk and sometimes an estimation task, a brief introduction to a new task, or a review of an old task. Remember to introduce new activities gradually, replacing the familiar activities with new ones over time.
3. **Remind the class of what they will be working on at the independent stations.**
   If a particular problem came up the day before, remind them of what you expect to see that day. If things went well that day, remind them of that.
4. **Excuse the children a few at a time to choose where they are going to work.**
5. **Watch for a few minutes to make sure the children have the materials they need, know what they are supposed to do, and are settled in.**
6. **Go around the room and observe and interact with the children.**
   - Make note of any needs you will have to handle later with the whole class or in a small group.
   - Don't expect it to be perfect. The environment you want to establish takes time.
   - Allow children to work with any set of activities for at least four days before you decide it's not working. It takes time for them to settle in with something new.
7. **After work time is over, ask the children to clean up and gather back on the rug.**
8. **Spend a minute or two reviewing how the math time went.**

7 The Concept Development Stations

What procedures need to be in place?

How can we make sure the children are learning at the stations?

The Concept Development Stations

Introduce the tasks as problems to be solved, rather than tasks to be completed. Our goal is to maintain a high level of involvement.

What Procedures Need to Be In Place?

Whenever you provide children the opportunity to choose where they are going to work, you must make sure that they are getting the experiences they need, no matter which activities they choose. This means you need to provide a set of activities that all deal with the same concept. For example, if the children are working with addition and subtraction, all the choices would give them various ways to practice addition and subtraction.

Establish the Following Procedures

The underlying structure of the independent work time is much like the structure used during self-directed exploration, so similar procedures need to be in place.

- **Introduce the activities to the whole class.**
  Introduce each activity to the whole class before putting it out at a station where children will be working independently. Introduce the activities over a period of several days, gradually replacing familiar activities with new ones until eight to ten tasks dealing with the same concept are available for the children to choose from.

  The way you present the tasks to the children will make a big difference in how they think about them. Since we are trying to establish an attitude of active participation rather than one of passively following directions, introduce the tasks as questions to be answered or problems to be solved rather than tasks to be completed.

  There are very subtle differences in these two approaches to presenting the tasks: one way encourages thinking and the other discourages it. For example, a teacher might introduce making geoboard designs by saying, "Your job is to make geoboard designs that have seven sides. Make sure you count the sides carefully and then record them on the paper." Or, the teacher could turn it into a question by saying instead, "How many different ways can you make shapes with seven sides? It will help you to keep track of all the ways you find if you make records of each of the ways as you find them." The first way of presenting the task leads the children to focus on getting the task done, while with the second way, the children are encouraged to find as many different ways as they can to complete the activity, prompting greater thought and creativity.

Let's consider another example. Measuring the number of scoops that could fit in a jar can also be introduced in two different ways. One approach focuses on how to do the task: "Here are the jars. You are to fill each one with scoops of rice and record how many scoops each jar holds." By introducing the task another way, the teacher can involve the children in thinking by asking the children to estimate first: "How many scoops of rice do you think this jar will hold?" After measuring out a few scoops, the teacher could then ask the children to consider the information gathered thus far and make adjustments: "Now that we see that three scoops filled the jar up to here, does anyone want to change their mind about the number of scoops the jar will hold?"

Introducing the task in this way models the thought process the teacher wants the children to use when doing the task on their own. The goal is for the children to continue asking these questions for themselves even when working independently.

When the directions for doing the activities are simple, they need very little introduction. It is important not to over-introduce the tasks so they will remain intriguing for the children. If the directions are somewhat complicated, take a few minutes at the beginning of math time for a period of two or three days to go over the instructions. Put the activity out at an independent station when most of the children know what to do. Don't worry if there are a few who do not know exactly how to do the tasks. The other children will be able to help them. Encourage them to help each other rather than to come to you for further directions.

When you have new tasks out, spend a day or two watching how the children work with them so you can redirect them (or reteach the activity to the whole class if necessary).

- **Have the children set up.**
If you have the materials for each activity organized in tubs or baggies, you can have the children set up the stations. I generally prepare station materials to accommodate six children at a time. The particular materials needed for the task, such as task cards and recording sheets, are put in baggies or tubs, and other basic materials, such as pattern blocks, toothpicks, or geoboards, can be added to the station by the children.

- **Let the children choose where to work and move from place to place whenever they are ready to work with something else.**
Our goal is to maintain a high level of involvement. Just as when children were exploring materials, the level of participation will be greater if the child makes the selection. If you have provided a set of activities that all deal with a particular concept, children can choose the task they want to work with and still be getting the experiences they need. For example, if all of the activities deal with counting, it doesn't matter which counting task the child works with. Or if all the tasks deal with place value, the child will be getting appropriate experiences no matter which station is selected.

  Choosing where to work does not imply that one can also choose not to work. On the contrary, the expectation is that a child will work harder because they have selected what to work with. Just as was true during self-directed exploration, if a child cannot make a choice and tends to wander, you can offer a narrower range of choices. You might say something like, "Do you want to make creations with the lids or do you want to record pattern block designs?"

- **Allow children to come back to the same activity as often as they wish.**
One of the advantages of an independent work time is that it provides opportunities for children to get lots of practice. Children learn from the repetition of tasks. Some teachers have felt children were more accountable if they kept a record sheet of where they had gone each day. While it may seem like this would help them to be more accountable, what actually happens is that they begin to focus on getting the tasks done and checking them off rather than getting engaged in the work itself.

It is not necessary for you or the children to keep track of where they have worked. It is more important to make sure the children are productively engaged and that you see signs of their growth as they work with whatever task they have chosen.

- **Allow children to work alone or with others.**
Children will naturally work together with those who have chosen the same tasks. Talking while they work can be an important part of their learning. Even though there often will be tasks designed to be done with partners, I do not recommend assigning partners. I allow the children to find partners when and if they happen to choose a partner game. If a child cannot find a partner, they will often just play the partner game alone. Requiring children to work in pairs during a time in which children must decide where and how long to work can create problems that are difficult for them to solve at this age. This can distract them from the mathematical work they are to be doing.

- **Remind the children that they don't need to come and get you.**
While you want to be very involved in what the children are doing as they work independently, you also want to foster independence so you can work with individuals and small groups when appropriate. Remind the children that they don't need to come and get you to show you what they are working on. Encourage them to talk to the other children if a problem arises to see if they can solve it before bringing it to the teacher. If there are persistent problems that require help in solving, then you need to deal with them during a whole class time so all the children know how to cope with the particular situation.

Get to Know Your Children and Respond to Their Needs

Observing the children while they work is one of our most important jobs. Once the stations are introduced and the children know what to do, you can focus on what they are doing. How the children do the tasks will become as important and as revealing as what they do. As you observe them, you will be able to see whether some need more instruction, whether the activities are too hard or too easy, or whether there is a common misunderstanding you need to deal with. You can also interact with the children, providing support and presenting challenges. If you have parent helpers or assistants in the room, assist them to learn first to observe the children to see how they work with the task before they intervene. It is important that all the adults in the room know it is important to intervene only when necessary, and then only to help the children work on their own.

Intervene if Necessary

Sometimes children say they are bored when they mean they want some adult attention. They may not be used to taking responsibility for themselves. They could be having trouble making a choice, or perhaps some are having difficulty accepting the fact that they didn't get their first choice. Other children just need to be redirected. You may say something like, "I would like you to choose a task and I will come by later and see how you are doing." Once in a while, you may find it helpful to work with a child for a few minutes on a task they choose. You can check to make sure the task is not too hard, and you can help the child see that they can get engaged in the task.

Have the Stations Available for Several Weeks

One of the keys to this approach, and one of the hardest aspects for many teachers to accept is the idea that the same activities should be available for children to work with for several weeks. In an environment that supports concept development, children are allowed the time they need to become proficient with these ideas. When new activities are introduced, it takes a few days for children to become deeply involved. Just knowing how to do an activity is the first step for children. Expect that they will become more rather than less involved after a few days. It takes a few days to "get into it."

When the tasks are appropriate, you will see the children working hard, making a few mistakes, double-checking to see if they did it right. After several days or even weeks, you will see less intensity as the tasks become easier for them. You will begin to notice less concentration and sense a little more restlessness. When you ask questions and they know the answers quickly without having to figure them out, then it is time to move on.

The Concept Development Stations

Occasionally Present Children with Assigned Tasks

On occasion, you will want to assign particular tasks for children to do rather than having them choose. Assigned tasks do not generally meet the range of needs in the same way that providing choices does. However, there may be times when assigning tasks is appropriate. There could be a particular activity or set of activities that everyone needs to experience, and so allowing choices would not be applicable. This would be particularly important if each of the tasks focuses on a different topic or strand.

There may also be times when you will want to give children assignments that require a minimum of movement and decision-making on their part so you can focus your attention more fully on a small group or individual assessments.

The following is an example of a set of activities that the teacher wants all her children to work with because it will give them experiences in different areas of the math curriculum. She has divided her class into four groups and assigned them the following tasks.

At one table, the teacher has asked the children to sort various collections in as many ways as they can.

At the next table, the teacher is giving the children the experiences they need to prepare them for later, more structured work with toothpicks. The children are creating toothpick pictures using their own ideas.

At a different table, the children are copying geoboard designs.

At still another table, children are making circles by tracing around lids. They then cut these circles into halves and fourths and make border patterns with these shapes.

Even when I have various stations I want all the children to work with, there are several reasons why I would still not rotate the children through the stations. From a practical classroom management perspective, when we set up stations for children to rotate through, we create a situation in which we need to find activities that can fit into a certain time slot rather than activities that meet a certain need.

Children do not work or learn at the same pace nor stay interested and focused for the same amount of time. Some children will invariably feel rushed and others will feel restless and bored.

When children are directed to move from activity to activity at a given signal, the children are not asked to assume the responsibility for learning or even for using their time effectively. Rather, the teacher is in control and the emphasis in on doing the activity in the time allotted rather than on learning a concept, developing a skill, or becoming truly engaged in the task. The motivation becomes external rather than internal, and the teacher's job can become one of keeping the children on task and/or taking care of discipline problems.

Instead of a rotation, I have used what I call a flip flop. I divide the children into four heterogeneous groups and assign each group a task. I want to allow for different finishing times, so I schedule an unstructured time (such as snack time, recess, free reading or working with chalkboards) following the work time. Then, at the beginning of another period (after recess or even the next day), children will go on to a different task. The key is that children have a place to go when they finish rather than moving to one of the assigned tasks. This allows children who work at different speeds to finish at various times.

How Can We Make Sure the Children Are Learning at the Stations?

When we make a commitment to ensuring success for all our students, we must find ways for every child to be productively engaged in appropriate math experiences. That means giving them time to work at independent stations where each child can engage in the tasks at their own pace. This also assumes they know how to do this, thus the need for learning to be self-directed learners. It really is the most natural way for children to learn but they need to know what that means. And we need to know how to get the best effort and engagement from the children.

You will know the independent work time stations are operating well when every child is working at their edge and developing competency with the concepts. You, the teacher, are there to make sure this is happening–sometimes by observing and identifying needs, sometimes by focusing the child's attention on a particular idea or relationship. The goal is for you to set things in motion and to move away so the children can be doing the thinking and learning. Of course, that requires constant work on your part just because of the number of children and the range of needs, but you should in some ways be invisible to the child who is working to figure something out. They should be almost unaware that they have been challenged or supported by your efforts. The role of independent work isn't to free the teacher to work with children but to give children the ultimate learning time: internal and personal. You must be a master at creating this situation–but not the center.

Make Sure the Children Know They Are Accountable for Working Hard. When children are allowed to move from station to station, they need to understand that this does not mean they are allowed to wander or waste time. Since many activities are open-ended and do not have recording sheets to go along with them, it is not obvious when a child has "finished their work." Some children will show evidence of more than you expected, and others will have less to show for their time with the task.

The children will decide for themselves when to move on to another task. What you need to do is to check to make sure each child is working hard, regardless of whether they seem to accomplish a lot or a little.

If some children are wandering or wasting time, you can set goals for them and come back to see if they have accomplished the goal. For example, you might say, "I would like to see if you can make a pattern all the way to the edge of the rug. I will come back later and see how you did."

What to Do When Problems Arise

1. **Treat problems that arise as learning opportunities rather than as interruptions.**
   They really do provide opportunities for increased understanding for both student and teacher!

2. **Be honest and direct with the children.**
   Remind them of your expectations. Don't assume that solutions or alternative behaviors are obvious to them. Discuss ways they might handle particular problems that may come up by posing, for example, the following kinds of questions.
   - What can you do if someone takes the place you wanted to work?
   - What can you do if you run out of the materials you need?
   - What can you do if someone knocks over your work?

3. **Don't adapt the rules for the whole class just to accommodate one or two children who aren't able to work independently.**

4. **Provide more guidelines for particular individuals if necessary.**

5. **Expect imperfection.**
   Instead of assuming everything will be textbook perfect before you try it, have an escape plan in mind. For example, if everything falls apart because you introduced an activity too quickly and no one knows what to do, or you don't have enough materials after all, or the game doesn't work the way you thought it would and everyone is confused – don't panic. Just have the children clean up the math materials and then read a story or sing a song. You can reintroduce the task or choose another one tomorrow.

Dealing with Common Problems and Concerns

Some problems are predictable as they exemplify why this time is needed. The following are situations that teachers often encounter.

1. **Some children can't make a choice. They wander around and waste time.**
   Some children need to check out all the choices first before really getting involved. If, after a day or two, they are still having difficulty making a choice, offer some help. You may limit the choices for them by asking, "Would you rather do this one or that one?" Just asking the questions may help them decide on one you haven't suggested. If, on the other hand, they are still unable to make a choice in a reasonable amount of time, you may need to let them know that they are responsible for working hard and that you will make a choice for them if they can't do it for themselves. Give them a chance the next day to choose by themselves again.

2. **Some children go to the same activity all the time. I don't think they are getting the experiences they need.**
   Children often know better than we do what they need to do, and they may be getting the experiences they need by repeating a particular activity. However, some children choose one particular activity over and over again because they feel comfortable there and know they can do what is expected. If you think the child needs to be encouraged to try something new, you might invite them to pick something that the two of you could try together.

3. **Some children seem to associate choosing with free play and don't feel accountable.**
   We commonly assume that children understand what we expect and are deliberately being uncooperative. It may be that we need to make our expectations more clear to them. Remind the children that their job is to work hard and that you are allowing them to choose so that they can find the job they are ready to work hard on. You need to let them know that you view this time as serious work time and not as a free play time.

4. **Some children get upset because they don't get their first choice. Some race to get a place or refuse to choose because they didn't get to do what they wanted.**
   Often, children simply need reassurance that the choices will be there again tomorrow and that there will be other opportunities to get their first choice. Over time, children will see for themselves that they don't have to fight to get what they want. It helps children choose activities calmly if you dismiss a few children at a time to choose where they are going to work. Remind them that they should have an idea about where they want to work before they are dismissed, because you will not wait until they have picked a place before calling others to choose. Sometimes it helps to have the children "role play" the process of choosing so they have a clear idea of what is expected.

5. **I try to provide a range of choices, but some children choose tasks that are too easy or too hard.**
   I have found that if I provide a range of choices, most of the children will choose the appropriate level. That allows me to focus on those few children who need more help or who need a challenge. After observing for a while to see how the child is working, I can determine whether the task is indeed too hard or too easy. I can then provide guidance if necessary.

6. **I just don't think kids are able to decide for themselves what level to work on. Many of my children will just do what is easiest. Other children will try to do things that are too hard just to keep up with their friends.**
   In my experience, children are able to decide for themselves what they need to work on if the classroom environment supports that expectation. However, in a classroom where children feel they are ranked or in competition with other children, some may respond by trying to do things that are too hard and others by giving up and doing whatever is easy. However in an environment that values and supports each child's growth, this doesn't become the

motivator. The teacher has an important role, not only in creating this environment, but also in observing children's needs so she can provide a challenge or help when necessary. Using "expandable" tasks makes this job easier than it would be if you had to provide a variety of tasks for a variety of needs.

7. **There is so much to deal with, I can't keep track of what everybody is doing.**
   I have found that if I try to watch too many children at once, I don't really see anything. You will be farther ahead in the long run if you observe just a few children closely. What you will find is that what you observe one child doing, you will notice others also doing. There really aren't 30 children doing 30 different things. There are clusters of children working in pretty much the same way. You won't be able to (nor do you need to) take notes on every child every day. Just make sure you focused in on a few children each day. Achieving that will help you guide the class as a whole.

   The particular individuals who really need you will make themselves known. This approach works because you are providing a variety of tasks that naturally meets a diversity of needs. These activities are available for children to experience over and over. The information you get from watching the children lets you know if there is some confusion you need to deal with, or perhaps a challenge you need to present. Know that over time, you will get better and better at picking up the clues from the children as you continue to work in this way.

8

Setting Up The Classroom

How do we get our classrooms ready?

Setting Up The Classroom

Children are more creative when they aren't afraid they will run out of materials. It's better to have lots of a few materials than small amounts of many different ones.

How Do We Get Our Classrooms Ready?

One of the first steps to take when creating an environment for learning mathematics is to make sure you have a variety of manipulatives available for the children to work with. To understand the mathematics they are working with, children need to be actively involved with materials. These materials have value not just as counters to help children get answers, but they allow mathematical ideas and relationships to be modeled. They provide children with tools for thinking and for testing their ideas.

Gather a Variety of Materials

It is important to provide children with several different kinds of manipulatives so they won't associate an idea with a particular object. There is not one "best" set of materials, but the following is a list that has proved to be more than satisfactory. It is not necessary to have everything on the list in order to get started. However, if you do have the following materials available, you will be able to provide a rich mathematics program for your students.

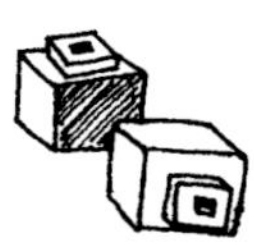

Unifix cubes *(1000-2000)*

Tiles *(800)*

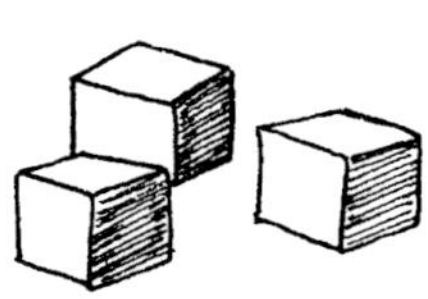

Wooden cubes *(400-800)*

Treasure Tub collections
(15-20 collections)

Each collection consists of 60-100 small items such as buttons, shells, screws, pebbles, bread tags, etc.

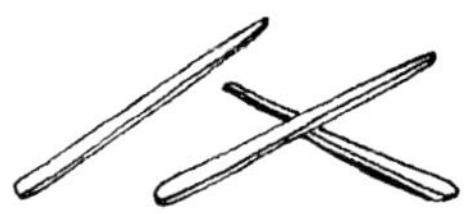

Toothpicks *(not round) 3-4 boxes*

Pattern blocks *(3 sets)*

Beans and portion cups *(500)*

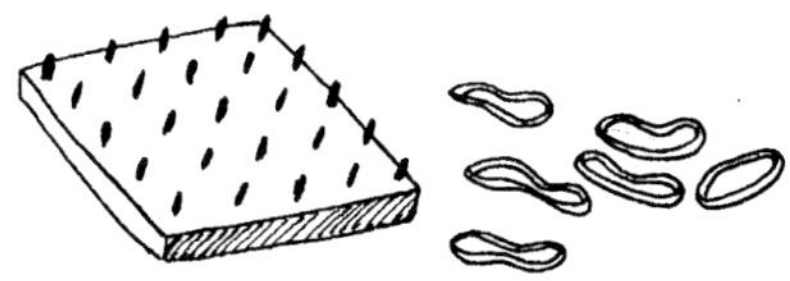

Geoboards/Geobands *(16)*

Individual whiteboards

Containers
(various items such as boxes, lids, small food containers, etc.)

Scales and objects to weigh, such as rocks, apples, oranges, potatoes, clay, small cans of food, balls, small containers of things (ex: beans, rice, popcorn, or sand.)

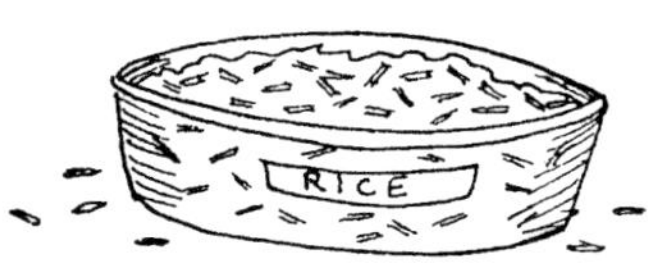

Rice and jars: *1 or 2 tubs of rice (or other pourable material such as bird seed or millet); jars of various sizes and shapes, labeled with A, B, C, etc. Various scoops and funnels.*

Setting Up The Classroom

These basic materials will be used for many activities. They should be stored in a place where children have access to them so they can help deliver them to the appropriate places in the room and put them away at the end of the work time. You can make your own shelves using large concrete bricks and 12-inch-deep boards, if shelving is not available.

No matter what math program you will be using, a variety of supplemental materials such as task cards, recording sheets, and game boards will be necessary. Group, label and store these according to the concepts for which they will be used to develop understanding.

Arrange the Room to Allow for Active Learning

You will need to set up the classroom to accommodate a variety of ways for the children to be actively involved. The room layout should allow for three basic settings: the whole class working together, independent stations, and teacher-directed small group work. When I am setting up a classroom, the first thing I try to create is a large area where the whole class can meet on the floor. There are several reasons why this is important. First of all, young children need to be close to the teacher to feel they are a part of the discussion. It is harder for some children to focus on what is happening if the teacher seems far away. (This meeting place also works well for story time, whole group language lessons, and music and movement activities.) A second reason for creating a large floor space is so the entire class can work together with materials on the floor. When everyone is busy on the floor, they can see what each other has built or created. Working on the floor naturally lets children engage with each other and gives them some flexibility in using space. Also, when children are sitting in the large group area, the tables and small floor areas are free so that independent activities can be set up around the room. The children can then easily move from the meeting place on the rug to choose where they want to work. In the past, I have had to set up classrooms where I had 33 children with 33 desks plus extra tables, so I know this can be a challenge. What helped me the most in setting up classrooms was finally realizing I could get rid of some of the furniture. As a teacher who had to work hard over the years to get the furniture I needed, it was difficult to give up the furniture I had, but it certainly made it easier to set up my room once I decided it was all right to do that.

Setting Up The Classroom

After I have established a large meeting area, I arrange the furniture to create the independent work areas. The desks and/or tables that the children use when doing work in other subject areas can double as math stations.

I make the whole room available for all projects. Math materials are stored in one place, writing materials and science materials are stored in other places, but the children work anywhere in the room.

Children need table or floor space to work with materials.
Tables work nicely, but if tables are not available, then desks can be pushed together to create table-top surfaces. Pushing the desks together also helps create room for that large group meeting place. I was able to provide work space for many children by using a piece of plywood as a table. I set the table top on large kindergarten blocks and the children sat on the floor to work. Up to ten children could sit around this table, and I didn't have to worry about chairs taking up space.

Setting Up The Classroom

It has worked best for me to have the tables (or clustered desks) serve as work areas for all of the children rather than as personal spaces. The children did not have assigned places at the tables or desks and stored their personal things in "cubbies." This was beneficial in that children could move from place to place in the room without a sense of being in other people's desks. We shared pencils, scissors, crayons, glue, and other supplies which were stored on a supply shelf. The children had a work folder to store their work in. I also used a "Finished Work" box and an "Unfinished Work" box.

Not all teachers choose to set up their rooms the way I did. Some prefer to have children work at their assigned desks during other times of the day, but use the children's desks as math stations during math time.

When you work with a small group, make sure you are facing the rest of the class so you can easily keep track of what is happening.

Setting Up The Classroom

I didn't need to set up a special place for working with a small group, as I could work with a small group either in the large group rug area or at one of the tables.

The key to setting up your room is to make use of every bit of space. The following page shows an example of how teachers have set up their rooms.

Here is just one example of a classroom arrangement. While the room is not large, the teacher has found ways to accommodate all the various classroom activities.

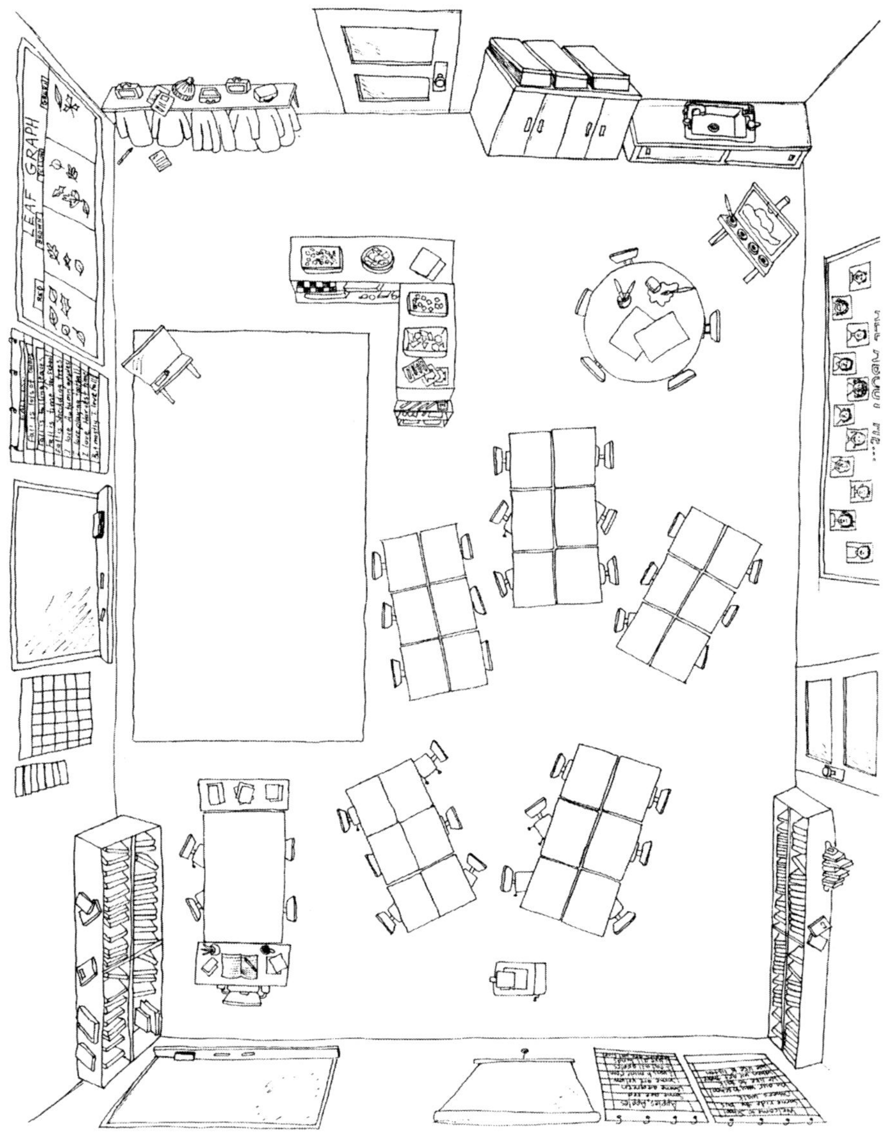

Room Arrangement

Setting Up The Classroom

Steps for Getting Started

Remember: You can't figure everything out ahead of time. So it's important simply to get started. Then you can look at where you are and consider what to do next.

1. **Gather the basic materials.**
   You don't need everything, but you do have to have enough materials to engage your class. Unifix cubes, toothpicks, and collections of things like shells, rocks and buttons can get you started.
2. **Plan storage for the materials.**
   You can create your own shelves with bricks and boards if necessary.
3. **Arrange your room to support an activity-based classroom.**
   Create a floor area large enough for the whole class to meet. Arrange tables or desks to make work areas for using materials.
4. **Plan activities and prepare the materials for a four- to six-week block of time.**
   The first block of time needs to be spent with self-directed exploration. After that, you will choose a particular concept to work on.
5. **Focus on the concept you are planning to teach.**
   What kinds of activities will you present?
6. **Consider how you are going to organize your children for instruction.**
   - Whole Group: Teacher-Directed Activities
   - Independent Station Time
   - Assigned Tasks
   - Small Group: Teacher-Directed Activities
7. **Begin.**
   Give yourself permission to learn.
8. **Reflect.**
   - What went well? What did not go well?
   - What would you do the same again?
   - What would you change?

9

It Takes Time

Our top priority is the development of children's understanding

Give children time to learn

Give yourself time to learn

Teach in ways that allow all children to be successful

It Takes Time

In a world where children are hurried from place to place, we need to create an environment where they are given time and are allowed to focus, to think, to contemplate, and to learn.

Our Top Priority is the Development of Children's Understanding

LEARNING

- Evolves and comes from many experiences
- Can result in partial understanding – even confusion is natural
- Is unique for each individual

STUDENTS

- Need to verify their thinking for themselves through direct personal experiences (often through the use of models)
- Should interact, discuss, verbalize (use language to communicate, not memorize)
- Must work together
- Have to know that it's okay not to know yet

TEACHERS

- Must be clear about the concepts they want to teach
- Need to be interested in what students are really thinking and understanding
- Should use models and activities to aid children's thinking about mathematical ideas
- Need to stimulate thinking by posing problems, asking questions, setting up situations, and focusing and refocusing the student's attention

Give Children Time to Learn

When planning mathematics (as well as other curriculum areas), we must take care to ensure we are making the best possible use of our time. That doesn't mean we should rush and try to fit too much in. In a world where children are hurried from place to place, we need to create an environment where they are given time and are allowed to focus, to think, to contemplate, and to learn. We do not want to underestimate the complexity of the ideas children must grapple with in developing an understanding of mathematical ideas. Children often need more time to work with an idea than we provide them. When we move children too quickly or too soon through experiences, they don't internalize or "own" the concepts they have been exposed to. Consider the numbers of fourth- and fifth-graders who are still counting on their fingers when asked such questions as "What is 14 + 9?" or who work diligently to apply learned procedures to solve problems like 298 + 42. These problems could be simple for them if they had been given the time it takes to develop real facility with numbers and number relationships.

Remember, we are building the foundation, so giving children time to develop competence is important. We will never have as much time as we might want to work with mathematics. What we must do is choose carefully how we spend the time we do have.

In the long run, it is much more effective not to rush children through mathematical tasks and not to assume that simply exposing them is sufficient. They need meaningful experiences that will enable them to work with the ideas at some level of understanding (even if they don't understand fully). We do not want children working on ideas that are out of reach, and thereby result in their success only for as long as they can remember what the teacher told them to do. We need to respect the real work of young children and not have them doing tasks that merely look challenging and advanced, but in reality are meaningless to them.

Trying to fit it all in should not be the goal. Choosing carefully and then giving children enough time to learn is the foundation on which good planning must rest.

Give Yourself Time to Learn

It is really possible to create a positive and productive learning environment that meets the needs of young children. There will still be challenges, difficult situations, and unexpected problems. Not every day will go smoothly. However, I have found that the challenges are more easily met in an environment that truly is respectful of and consistent with how young children learn. Some of the unexpected moments will be quite wonderful: when the children groan because it is time to go out to recess and they would rather work on math; or when a child giggles with delight because he was able to accomplish a task that once seemed way too hard; or even when the children clean up in less time than you were planning. (It really can happen!)

Teach in Ways that Allow All Children to Be Successful

When creating the environment for learning mathematics, our goal is for all children to develop mathematical concepts in an environment that acknowledges and values each child's efforts to grow and to learn. In this environment you will see children working hard and working independently. You will see children engaged in tasks that have meaning for them. The children go beyond what you might expect. They have confidence in their own abilities to figure things out. This approach works and is effective because it makes sense to children. It engages them in thinking and working hard. It gives them opportunities to practice and to grow. This approach works because everyone can be successful.

References

Myren, Christina. *Posing Open-ended Questions in the Primary Math Classroom.* Bellingham, WA: Math Perspectives, 2013.

Richardson, Kathy. *Developing Number Concepts: Planning Guide.* Parsippany, NJ: Dale Seymour Publications, 1999.

Richardson, Kathy. *Developing Number Concepts Book 1: Counting, Comparing, and Pattern.* Parsippany, NJ: Dale Seymour Publications, 1999.

Richardson, Kathy. *Developing Number Concepts Book 2: Addition and Subtraction.* Parsippany, NJ: Dale Seymour Publications, 1999.

Richardson, Kathy. *Developing Number Concepts Book 3: Place Value, Multiplication and Division.* Parsippany, NJ: Dale Seymour Publications, 1999.

Richardson, Kathy. *Developing Number Concepts in Pre-Kindergarten: Children Learn Number Concepts: A Guide to the Critical Learning Phases.* Bellingham, WA: Math Perspectives, 2008.

Richardson, Kathy. *Understanding Geometry.* Bellingham, WA: Math Perspectives, 1999.

NOTES

NOTES

Math Perspectives Teacher Development Center publishes professional resources and provides long-term professional development and courses for teachers of grades PK-6 mathematics. To learn more about Math Perspectives wide range of professional development visit mathperspectives.com or call 360-715-2782.

MATH PERSPECTIVES Best-selling Resources
by Kathy Richardson

– *Developing Math Concepts in Pre-Kindergarten*

– *How Children Learn Number Concepts: A Guide to the Critical Learning Phases*

– *Developing Number Concepts Series*
  - Book 1: Counting, Comparing and Number
  - Book 2: Addition and Subtraction
  - Book 3: Place Value, Multiplication and Division

– *Assessing Math Concepts Series*
  - Assessing Math Concepts in Pre-Kindergarten
  - Assessing Math Concepts in Grades K-2

– *Understanding Numbers Series*
  - Place Value
  - Addition and Subtraction
  - Decimals

– *Posing Open-Ended Questions in the Primary Math Classroom by Christina Myren*